Palgrave Studies in Impact Finance

Series Editor
Mario La Torre, Department of Management,
Sapienza University of Rome, Rome, Italy

The *Palgrave Studies in Impact Finance* series provides a valuable scientific 'hub' for researchers, professionals and policy makers involved in Impact finance and related topics. It includes studies in the social, political, environmental and ethical impact of finance, exploring all aspects of impact finance and socially responsible investment, including policy issues, financial instruments, markets and clients, standards, regulations and financial management, with a particular focus on impact investments and microfinance.

Titles feature the most recent empirical analysis with a theoretical approach, including up to date and innovative studies that cover issues which impact finance and society globally.

Wolfgang Spiess-Knafl

The Sustainable Finance Operating System

The Value Chain of Capital

palgrave
macmillan

Wolfgang Spiess-Knafl
European Center for Social Finance
Munich Business School
Munich, Germany

ISSN 2662-5105 ISSN 2662-5113 (electronic)
Palgrave Studies in Impact Finance
ISBN 978-3-031-97498-4 ISBN 978-3-031-97499-1 (eBook)
https://doi.org/10.1007/978-3-031-97499-1

This Palgrave Macmillan imprint is published by the registered company Springer Nature Switzerland AG
The registered company address is: Gewerbestrasse 11, 6330 Cham, Switzerland

Competing Interests The author has no competing interests to declare that are relevant to the content of this manuscript.

Contents

List of Figures

List of Tables

CHAPTER 1

Setting the Stage: The Why and What of Sustainable Finance

Abstract This chapter introduces sustainable finance, defining its core concepts and outlining the necessity for integrating environmental, social, and governance (ESG) factors into a financial system facing urgent climate and societal challenges. Critiquing the traditional system's limitations in handling externalities and information gaps, it explores the key drivers for change, emphasizing the economic imperatives of climate action ("E") and social stability ("S"). The chapter examines the distinction between ESG perspectives focused on financial "value" versus ethical "values," while also addressing the challenge of greenwashing and the regulatory efforts to ensure authenticity in sustainability claims. Contrasting the rules-based EU regulatory approach with the norms-based US model, the chapter analyzes their differing philosophies and implications for global market fragmentation. Finally, it introduces the "Sustainable Finance Operating System" as a conceptual framework that structures the exploration of how capital is guided toward sustainability goals.

Keywords ESG (Environmental Social Governance) · Climate change · Greenwashing · Sustainable finance architecture · Regulatory frameworks

W. Spiess-Knafl, *The Sustainable Finance Operating System*, Palgrave Studies in Impact Finance,
https://doi.org/10.1007/978-3-031-97499-1_1

1.1 Defining the Landscape

A Changing Financial Environment

The global financial system is currently undergoing a transformation as it is confronted with the challenges and long-term costs of climate change, social inequality, and environmental degradation (Migliorelli & Dessertine, 2020). Investors, regulators, and companies are recognizing that traditional financial metrics alone cannot capture the full spectrum of risks and opportunities (Schoenmaker & Schramade, 2018). This book explores how environmental, social and governance (ESG) considerations are integrated into financial decision-making, risk management and investment strategies.

In recent years, sustainable finance has moved from the periphery of academic and policy debates into the mainstream (Busch et al., 2021a, 2021b). Central banks are incorporating climate risks into their stress tests (Acharya et al., 2023). Regulatory bodies across Europe, the United States and beyond are instituting robust disclosure requirements (e.g., Hummel & Jobst, 2024). Financial institutions are deploying advanced data analytics to assess ESG performance (Boffo & Patalano, 2020). At the same time, public expectations and activist movements, from youth-led demonstrations to institutional divestment campaigns, are driving a shift toward greater transparency and accountability in how capital is allocated (Gunningham, 2020). This happens against the background of populist politicians in government positions and the politicization of ESG (Hilson, 2024).

Core Concepts and Pillars of Sustainable Finance

In recent years, finance has evolved beyond its traditional focus on risk and return to including broader societal and environmental objectives. Sustainable finance is an emerging paradigm that integrates environmental, social, and governance considerations into financial decision-making, ensuring that capital not only generates financial returns but also contributes to a more sustainable future.

At its core, sustainable finance is the practice of allocating capital in ways that support long-term economic growth while addressing environmental degradation, social inequality and governance issues using different strategies. It represents a shift from viewing investments solely through the lens of risk-adjusted profit optimization to considering their broader impact on society and the planet. In essence, sustainable finance aims to balance financial performance with the imperatives of responsible stewardship and social value (Schoenmaker & Schramade, 2018).

The framework of sustainable finance is built on three key pillars as outlined in Table 1.1. Environmental factors focus on reducing negative impacts on nature, such as lowering greenhouse gas emissions, preserving biodiversity, and conserving natural resources. Social considerations prioritize community well-being, including fair labor practices, health, and education. Governance examines the systems by which companies are directed and controlled, emphasizing transparency and ethical conduct. The analysis of issues related to governance has been on the agenda for decades and widely researched. Together, these pillars commonly referred to as ESG form the foundation for making informed, responsible investment decisions (Li et al., 2021).

These factors are often seen from a risk perspective, but the importance of sustainable finance goes beyond mere risk management. By encouraging companies through different mechanisms to adopt sustainable practices, it drives innovation and long-term resilience (e.g., Löffler et al., 2021). This also reduces the likelihood of crises that can stem from environmental disasters, social unrest, or poor governance (Battiston et al., 2017). In doing so, it helps to reallocate capital toward enterprises that are not only profitable but also contribute positively to society.

As the financial landscape continues to evolve, sustainable finance is becoming increasingly central to policy debates, investor strategies and corporate reporting. It challenges traditional financial models to incorporate broader societal values, paving the way for a financial system that is more transparent, inclusive and aligned with global sustainability goals. Yet, as evidenced by the attacks on ESG and DEI programs, this landscape remains evolving (Kishan & Brogger, 2025).

1.2 The Traditional Financial System: Need for an Upgrade

How Money Flows: The Intermediated Chain

Traditional finance operates through a complex chain of intermediaries. It facilitates capital flow, but inherently distances capital providers from the ultimate use of their funds. Usually, multiple intermediaries stand between the suppliers of capital such as individuals, pension funds, or

Table 1.1 ESG framework. *Source* Li et al. (2021)

Dimension	*Factors*	*Definition*
Environmental (E)	GHG emissions Energy consumption and efficiency Air pollutants Water usage and recycling Waste production and management (water, solid, hazardous) Impact and dependence on biodiversity Impact and dependence on ecosystems Innovation in environmentally friendly products and services	Environmental matters that may have a positive or negative impact on the financial performance or solvency of an entity, sovereign or individual
Social (S)	Workforce freedom of association	Social matters that may have a positive or negative impact on the financial performance or solvency of an entity, sovereign or individual
	Child labor Forced and compulsory labor Workplace health and safety Customer health and safety Discrimination, diversity and equal opportunity Poverty and community impact Supply chain management Training and education	

(continued)

Table 1.1 (continued)

Dimension	*Factors*	*Definition*
	Customer privacy Community impacts	
Governance (G)	Codes of conduct and business principles Accountability Transparency and disclosure Executive pay Board diversity and structure Bribery and corruption Stakeholder engagement and shareholder rights	Governance matters that may have a positive or negative impact on the financial performance or solvency of an entity, sovereign or individual

insurance companies and the entities that require funding including corporations, infrastructure projects, or small businesses (Mishkin, 2007). Figure 1.1 illustrates this multi-layered value chain, where an individual either deposits money in a bank, which then lends to businesses or invests through an asset manager that allocates capital into a fund or even a fund of funds.

Each intermediary specializes in a particular segment of the market. Banks are efficient in underwriting loans, while asset managers specialize in portfolio construction. This arrangement can reduce transaction costs

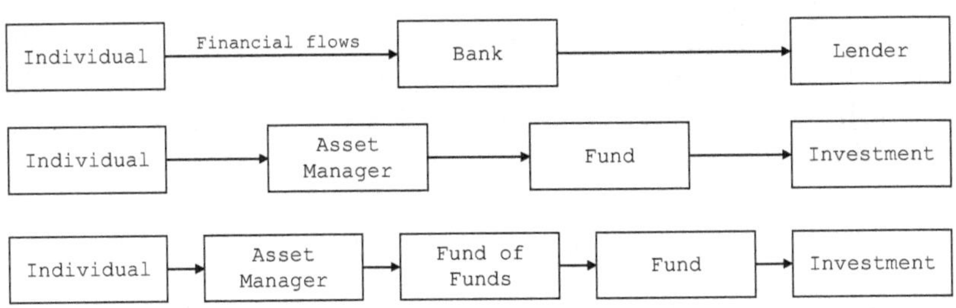

Fig. 1.1 The intermediation in the traditional financial markets. *Source* Own illustration

and leverage sector-specific expertise, effectively ensuring that capital is deployed in ways that maximize traditional financial performance (Levine, 2005).

Beyond Risk-Return: Limitations of the Traditional Model

Historically, the principal objective for all participants in this value chain has been to optimize risk-adjusted returns. For instance, a pension fund evaluating an asset manager primarily examines metrics such as financial performance, underlying risks or investment strategies. In turn, that asset manager chooses underlying funds or securities based on similar considerations (Elton & Gruber, 1997). Banks use credit scores and interest rate spreads to manage default risk while seeking to generate competitive interest income (Boyes et al., 1989). This focus on optimizing risk-adjusted financial returns inherently sidelines environmental and social considerations, treating them as external to core decision-making.

Blind Spots: Information Gaps in Traditional Finance

While this system is highly efficient for financial purposes, it tends to limit transparency regarding the ultimate deployment of capital. Intermediaries usually share only the essential data needed for the next entity in the chain such as overall returns or aggregated risk measures. They do not provide detailed disclosures on the specific nature of the underlying projects or borrowers.

A bank typically communicates general performance metrics to depositors, but not loan-by-loan details. Similarly, a fund-of-funds manager might only provide top-level information about portfolio composition and returns to investors, rather than in-depth insights into each investee company's operations. This opacity means capital suppliers historically lacked the necessary information to align their investments with broader sustainability values or assess non-financial risks embedded in the underlying assets.

Narrow Efficiency: Ignoring Externalities

Despite its limitations, this specialized structure offers tangible benefits. By focusing on risk and return, each intermediary can leverage economies of scale and deploy sophisticated analytical tools to optimize financial outcomes. Banks and asset managers maintain streamlined operations, reducing administrative overhead for capital owners, who in turn benefit from professional risk management and diversified portfolios. The result is a market that allocates capital quickly and at relatively low cost. It also makes sense as the intermediary is responsible monitoring the deployment of capital (Diamond, 1984).

This setup is traditionally seen as an efficient solution for channeling resources toward their most financially productive uses. However, this financial efficiency ignores environmental and social externalities. Sustainable finance fundamentally challenges this narrow definition, arguing that the system must account for these broader impacts to ensure long-term value and system resilience.

The Regulatory Push: Reshaping the Market

In recent years, new regulations and standards related to sustainability have begun to reshape these established practices. Policies such as the EU Sustainable Finance Disclosure Regulation (SFDR) and various climate-related disclosure mandates require more granular reporting on environmental, social, and governance (ESG) metrics.

This regulatory pressure cascades through the value chain, forcing participants to move beyond purely financial metrics and actively assess the carbon footprint, social impact, and governance structures of their portfolios and loan books. As a result, each intermediary is increasingly pressured to assess not only the financial viability of a project but also its carbon footprint, social impact or governance structure. This shift is forcing capital providers to consider non-financial factors in addition to traditional risk-return profiles (Schoenmaker & Schramade, 2023).

1.3 Why are we Doing it: E for Climate Change

Global Consensus and Concern

There is a widespread public and scientific consensus that climate change is one of the most critical challenges of our time due to the inherent systemic risks. In the European Union, more than 77% of respondents consider climate change a very serious problem. The majority attribute responsibility for tackling the issue to public authorities and business sectors. Only 35% feel personally accountable. Importantly, over 85%

believe that both the EU and national governments must act decisively to boost energy efficiency by 2030 and 58% advocate accelerating the transition to renewable energy in light of recent energy price spikes and gas supply challenges (European Commission, 2023a).

Similarly, in the United States, recent survey results reveal a divided yet concerned public. Nearly two-thirds of Americans say that climate change is already affecting their local communities. While opinions differ along party preferences with roughly equal shares viewing climate policies as either economically beneficial or detrimental there is broad criticism of large corporations for not doing enough. Most Americans expect to adjust their lifestyles in response to climate change, underscoring a collective call for balanced policy measures to address this pressing issue (Kennedy & Tyson, 2024).

Actions are necessary as climate change impacts our lives slowly but consistently. Through changing weather patterns, it not only impacts agricultural yields and the frequency of natural disasters, but also the development of urban heat islands among other things. In addition, there are cascading effects to consider. A good example are heavy rainfall as they lead to inflow of stormwater to wastewater systems, damage to property, potential high cost of evacuation as well as burden to public budgets among others (Lawrence et al., 2020).

The Economics of Climate Impact

It is possible to quantify damage caused by shifts in climate patterns, extreme weather events and gradual warming across regions with monetary values. Integrated assessment models perform this task by linking greenhouse gas concentrations to physical impacts such as sea level rise, altered ocean currents, more frequent extreme weather events, and gradual warming. These impacts are then translated into economic terms for sectors like agriculture, human health, infrastructure, ecosystem services, and tourism (e.g., Diaz & Moore, 2017; Söderholm, 2012).

Recent work updates key model inputs such as temperature response relationships, socioeconomic projections, and damage functions and arrives at an estimate of about US$185 per metric ton of CO_2 when using a two percent discount rate. Earlier central estimates often lay in the range of US$50 to US$70, reflecting differences in damage assumptions and socioeconomic scenarios (Rennert et al., 2022).

Estimates of the social cost of carbon are highly sensitive to methodological choices, particularly the selection of the discount rate, the treatment of low probability but high-impact outcomes and whether damages are counted globally or restricted to domestic effects. For example, applying a higher percent discount rate and limiting the analysis to domestic impacts led the U.S. Interagency Working Group in 2017 to adopt a much lower estimate of around US$12 per ton (Rennert et al., 2025). Such variation underscores the importance of transparent model documentation and the need to refine damage functions as empirical evidence advances.

Mitigation and Adaptation Strategies

Climate change is mainly driven by the emission of greenhouse gases (GHG), which include carbon dioxide, methane, nitrous oxide, and fluorinated gases. These gases are emitted through the burning of fossil fuel, deforestation, or through agricultural practices.

Consequently, one of the main strategies to mitigate climate change is to reduce these emissions. Thus, mitigation strategies focus on reducing the root cause of GHG emissions.

There are thousands of potential ways to reduce GHG emissions. It can include the insulation of buildings, less consumption of meat, less waste in the food supply chain, better use of old materials or increased use of bicycles to mention just a few. Overall, there are hundreds of thousands of potential concepts that might help to reduce overall GHG emissions. A good example is apartment blocks built in Eastern Europe some 50 years ago. They have a typical carbon footprint of 205 kWh per square meter and year. With advanced refurbishment you can bring it down to 92 kWh per year which shows the energy saving potential for one single area (Tabula & Episcope, 2017).

Adaptation strategies are focused on increasing the resilience of societies to better cope with the unavoidable effects of climate change. Defining and financing adaptation strategies is complex, given that impacts are slow-moving and have significant social dimensions, including threats to food security, health, and potential displacement or conflict.

It is also clear that there is a social dimension of climate change and its effects. In this regard, certain regions face different risks. While food security is more of a topic for emerging countries, health-related issues are a global concern. There are topics such as longer heat periods, increased danger of floods but also problems of forced displacement. Resource scarcity might also lead to conflicts over allocations (Otto et al., 2017).

The implementation of these strategies requires more than technologies which align well with sustainable finance approaches. The global food supply chain is a good example. Minimizing food loss and wastage could help to reduce emissions quite significantly and some of the solutions are straightforward. Farmers might benefit from better weather prediction data. In addition, transport systems can be upgraded and logistics made more efficient.

Funding the Transition

Keeping track of the various estimates for funding the green transition can be challenging. Deloitte (2023) estimates that the "transformation entails significant investments, on the order of US$5 trillion/year to more than US$7 trillion/year through 2050." The International Energy Agency (2021) estimates that to "reach zero emissions by 2050, annual clean energy investment worldwide will need to more than triple by 2030 to around $4 trillion."

The International Renewable Energy Agency (2023) states that "although global investment in energy transition technologies reached a new record of USD 1.3 trillion in 2022, yearly investments must more than quadruple to over USD 5 trillion to stay on the 1.5 °C pathway." The consultancy McKinsey (2022) even "estimates that the cumulative capital spending on physical assets for the net-zero transition—such as technology, infrastructure and natural resources—would need to change from today's annual average of $5.7 trillion to $9.2 trillion through 2050."

While the overall sums are daunting, individual projects often involve manageable investments. For example, a new EV charging station for residential use might cost as little as €1,000, solar panels around €20,000 and a fully automatic waste sorting machine up to €5 million. Typically, these specific project costs have an underlying business case to justify the investments.

1.4 Why are we Doing it: S for Societies Under Stress

Why Social Stability Matters

Just as environmental degradation poses systemic risks, social instability directly threatens economic prosperity and financial market stability. This relationship makes societal health a critical "S" factor demanding financial attention.

Societies around the globe are under stress from various sources. Democratic institutions are facing challenges as corruption, reduced transparency, and diminished accountability leave citizens feeling alienated and disempowered (Levitsky & Ziblatt, 2018). At the same time, widening income and wealth gaps undermine social cohesion (Piketty, 2018) Educational disparities continue to limit workforce development and innovation (Reardon, 2013). At the same time, the concentration of media ownership, coupled with the rise of misinformation, threatens the plurality and integrity of public discourse (Hindman, 2018). These issues, though diverse, are interconnected and have far-reaching implications for the overall stability and progress of societies.

The Erosion of Trust

The weakening of democratic institutions results in reduced public trust and hinder effective governance. When government processes lack transparency and accountability, citizens are less likely to participate in decision-making. As a result, public resources may be allocated inefficiently (Fung, 2015). Sustainable finance initiatives in the shape of impact investments can help bridge this gap by funding projects that enhance civic engagement and modernize local governance.

There are many impact pathways to achieve these outcomes. For example, growing disparities in income and wealth impose significant strain on social cohesion and economic stability. When a large segment of the population lacks access to essential services and opportunities, it undermines long-term growth and fosters social unrest (e.g., Berg et al., 2018). Instruments such as social impact bonds and social bonds have been designed to finance projects that improve affordable housing, healthcare, labor market integration and education in underserved areas. By targeting investments that address the root causes of inequality, these financial mechanisms contribute to a more balanced distribution of resources and foster stronger community resilience.

Another example is media freedom. A vibrant and diverse media landscape is essential for a healthy democracy, yet the consolidation of media ownership and the proliferation of misinformation have seriously undermined public debate and accountability (Neff & Pickard, 2024). Sustainable finance can prioritize platforms that ensure diverse and reliable information flows or provide funding independent media funds.

The Mechanics of Progress

There are different mechanics of progress, but they are best illustrated with education as access to quality education is a cornerstone of societal prosperity. Yet, many regions, particularly in emerging markets, continue to struggle with underinvestment in education, leading to significant skill gaps. These disparities not only hinder individual economic opportunities but also stifle broader economic innovation (Hanushek & Woessmann, 2012).

Figure 1.2 shows the underlying relationships between different Sustainable Development Goals (SDGs). Education reduces poverty by increasing income levels (Patrinos, 2024). Education also helps to enhance food security as it is an important driver of agricultural efficiency levels (e.g., Ogundari, 2014). Education was also found to have positive health outcomes (e.g., Psaki et al., 2019). Education can help women to overcome gender discrimination and thus contribute to gender equality (Vladimirova & Le Blanc, 2016).

1.5 Two Perspectives on ESG: Value Versus Values

Semantic Confusion

There is widespread confusion surrounding sustainable finance and associated terms like ESG (Environmental, Social and Governance), SRI (Socially Responsible Investing), and Impact Investing. It is based on the misunderstanding between value and values. Some investors and managers focus on value which is related to financial risk and return implications of ESG factors, while others focus on values such as non-pecuniary preferences, ethics, and the desire for real-world impact (Starks, 2023).

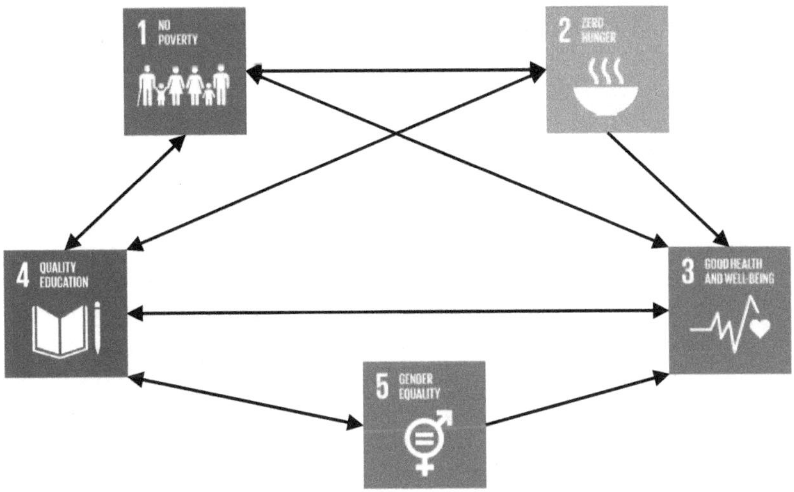

Fig. 1.2 The interconnectedness of SDGs. *Source* Own illustration

Sustainable finance strategies exist on a spectrum. "Value" investors incorporate ESG to manage risk (e.g., climate, reputation, regulatory) and to identify return opportunities. "Values" investors prioritize non-financial goals, often using negative screening (SRI) or seeking specific positive impacts (Impact Investing), sometimes accepting lower financial returns. Of course, there are approaches that blend both strategies.

The Values Lens: Investing for Real-World Impact

There is a wide range of individuals and asset managers which follow a strategy to achieve real-world outcomes with their investments. In private markets, they would use impact investment funds to fund renewable energy projects, education providers in emerging markets or climate tech start-ups (Spiess-Knafl & Scheck, 2023). In public markets, these investors would exclude certain sectors such as tobacco, gambling, or defense and focus on companies which are operating a green or social business model. A sign of this reallocation of capital is seen in the lower cost of equity for companies which are green or have a high ESG score (Gonçalves et al., 2022; Wang et al., 2023).

The rapid proliferation of sustainable investment products has significantly expanded the capital available for projects that meet these sustainability criteria. This reallocation of resources reflects the shifting values of investors and a growing preference for companies that integrate sustainability into their core business strategies.

The Value Lens: Risk Management and Opportunity

On the other side, investors using environmental, social, and governance considerations see it either as a way for identifying and mitigating risks or for identifying sustainable and growing long-term revenue streams.

Addressing environmental, social, and governance challenges proactively, companies can avoid costly regulatory fines, reputational damage, and operational disruptions. Oil spills, inappropriate governance structures, a track record of unethical behavior, risky supply chains including child labor, or heavy pollution are risks for the existence of companies. The main idea is that higher ESG scores indicate good risk control and compliance standards. Higher ESG scores also indicate that these companies have a lower frequency of incidents impacting the share price (Giese et al., 2019). ESG frameworks are also used outside of public markets by private equity funds (Zaccone & Pedrini, 2020) or less prominently for venture capital funds that invest in start-ups that usually have less ESG-related risks given the limited size of their operations (Lange & Banadaki, 2023).

The other perspective with this lens considers ESG as a framework that aligns companies with long-term business trends. This view holds that sustainable practices not only mitigate risks but also position companies to capitalize on fundamental shifts in the market. Renewable energy, energy efficiency, electric mobility, decarbonization but also education, healthcare, agriculture, and nutrition are all long-term growth trends. In addition, customers are often found to have a higher willingness to pay for sustainable products (e.g., Li & Kallas, 2021; Durán-Román et al., 2021).

1.6 Navigating the Pitfalls: The Challenge of Greenwashing

Defining Greenwashing

Sustainable finance has moved from the margins of the investment industry to the mainstream, with trillions of euros or dollars flowing into ESG-labeled funds and sustainability-linked financial products. Obviously, the increase in available capital has also led to ethical problems as greenwashing or impact washing started to occur. The main question for investors is focused on the use of proceeds and the trustworthiness of the recipient (e.g., Achleitner et al., 2013; Maxwell & Lévesque, 2014). Greenwashing refers to the practice of making exaggerated or misleading claims about environmental and social impact.

Greenwashing can be observed in multiple forms and different levels. One form is label-based greenwashing, where terms like "sustainable," "green," or "impact" are used without clear criteria or measurable outcomes. This form of label-based greenwashing can be found for products, services, and investment funds. There is also process-based greenwashing, in which companies may have ESG policies on paper but do not apply them consistently in their operations (de Freitas Netto et al., 2020; Santos et al., 2024).

For the investment sector, we can observe similar strategies. Scope-based greenwashing happens where only the most obvious unsustainable sectors, such as coal, are excluded, while other polluting industries remain part of the portfolio. Another form is data-based greenwashing, which occurs when selective or incomplete ESG data is relied upon to make investments appear more sustainable than they truly are (Dempere et al., 2024).

A major driver of greenwashing is the lack of standardization in ESG data. Most funds depend on third-party providers like MSCI, Sustainalytics or Bloomberg, each employing different methodologies. This situation leads to inconsistent ESG scores; for instance, one rating might remove Tesla from an ESG index while another keeps ExxonMobil on. Consequently, there is potential to game the system.

At its core, greenwashing is thus about marketing ESG credentials without delivering real impact. The lack of standardization allows asset managers to define sustainability on their own terms, often resulting in misleading claims.

New regulation is meant to combat misleading ESG claims. Unlike financial reporting, ESG disclosure rules are fragmented and inconsistent across jurisdictions with the European Union being a frontrunner. Regulators have so far adopted a "say-do" approach as discussed below in more detail: they do not define what ESG means but require that funds align their marketing claims with their actual investments.

For example, the EU's Sustainable Finance Disclosure Regulation (SFDR) classifies funds in a way that distinguishes between general funds and those with higher sustainability commitments, while the SEC in the United States focuses on misleading ESG claims and failures to implement stated sustainability policies. Similarly, emerging standards from bodies such as the International Sustainability Standards Board represent global efforts to unify ESG reporting.

Greenwashing in Practice

Several high-profile cases have revealed the extent of greenwashing in financial markets. One notable example involves WisdomTree Asset Management which had to pay $4 million over greenwashing charges (Reuters, 2024). The firm marketed ESG funds that purportedly excluded fossil fuels, yet investigations found that these funds held securities of companies involved in natural gas utilities, coal-related businesses and even entities with partial fossil fuel activities. The SEC statement also underlines the regulatory approach mentioned above which stated that "at a fundamental level, the federal securities laws enforce a straightforward proposition: investment advisers must do what they say and say what they do" (SEC, 2024b).

In the WisdomTree case, the firm relied on third-party data from different vendors with inconsistent screening methodologies. One vendor classified companies by primary business sector, leading to the exclusion of only those companies in the "Energy Sector," while companies in the "Utilities Sector" that distributed natural gas were not excluded. Another vendor provided data on various aspects of fossil fuel activities, but WisdomTree did not subscribe to the full range of data sets (SEC, 2024a).

The distinction between sector-based and activity-based screening further complicates matters, as some companies are classified as "green" despite having involvement in fossil fuel activities.

Carbon Market Fraud: The Dark Side of Offsetting

Another issue is fraud in the voluntary carbon markets. Carbon offsets are intended as a tool for companies and investors to claim sustainability credentials by compensating for their emissions with funding for projects that reduce carbon elsewhere. In theory, this mechanism should support genuine emissions reductions (Kreibich, 2024).

In practice, however, the carbon market has become susceptible to fraud and manipulation which have undermined the credibility of the market. Additionally, low-quality offsets from projects with inadequate monitoring have further eroded trust. Notable scandals in this area include investigations revealing that many rainforest carbon credits were essentially worthless. For example, one investigation revealed how South Pole, the largest carbon-offsetting firm, sold millions of credits from its flagship forest project in Zimbabwe based on massively overstated emissions reductions. An investigation detailed concerns over financial mismanagement, flawed verification by accreditor Verra and how corporations used these questionable credits for "carbon neutral" branding, exposing flaws in the voluntary carbon market (Blake, 2023).

1.7 EU vs US Approaches to Sustainable Finance: Divergent Paths

Setting the Scene: Rules vs. Norms in Sustainable Finance

As sustainable finance matures globally, two dominant yet contrasting regulatory paradigms have emerged: a comprehensive rules-based framework spearheaded by the European Union and a more market led, norms-based approach characteristic of the United States (Chiu, 2023).

These models reflect deeper differences in governance philosophies, with Europe's precautionary principle favoring prescriptive standards and societal alignment and the United States' emphasis on market efficiency and financial materiality driving flexibility and voluntary norms. This is also seen in observations such as the "Brussels effect" where the European Union is seen as a global rule-setter (Bradford, 2020).

These models are also shaped by distinct political landscapes, including recent shifts in US policy that have challenged ESG initiatives, as well as varying market structures and stakeholder priorities.

Within the Sustainable Finance Operating System, these contrasting approaches manifest most clearly in the rule setting layer, where the

EU prescribes detailed requirements while the US relies on overarching principles and in the language layer, where the EU taxonomy mandates specific definitions and the US favors a looser, market-driven vocabulary.

The European Union: A Rules-Based Regulatory Framework

The European Union has firmly positioned itself at the forefront of sustainable finance regulation by developing a suite of legally binding measures that aim to systematically redirect capital flows toward sustainability objectives. At the core of this design is the EU Sustainable Finance Action Plan, introduced in 2018, which set out a clear vision for integrating environmental and social considerations across the financial sector (European Commission, 2018). By articulating concrete goals such as channeling investment into green technologies, enhancing transparency and safeguarding against misleading claims, the Action Plan lays the foundation for the operating system rule setting layer, guiding all subsequent regulatory initiatives. Finance is thus framed as a means toward sustainability (Ahlström & Monciardini, 2022).

To give this framework practical effect, the EU adopted a series of interlocking regulations that together construct its configuration of both the rule setting layer and the language layer within the Sustainable Finance Operating System. The EU Taxonomy provides a detailed science-based classification system of activities deemed environmentally sustainable, creating a shared vocabulary that investors and companies must follow (European Commission, 2020). Building on this, the Sustainable Finance Disclosure Regulation (SFDR), which entered into application in March 2021, requires asset managers and financial advisers to publish mandatory disclosures on how they integrate sustainability risks and adverse impacts into their investment decisions, thus embedding precise reporting obligations into the system rule book (European Commission, 2019). The Corporate Sustainability Reporting Directive (CSRD) then vastly expands the scope of non-financial reporting by obliging around fifty thousand companies to adhere to common standards under the European Sustainability Reporting Standards, elevating the quality and comparability of the information flowing through the data and information layers (European Commission, 2022). Finally, the European Green Bond Standard offers a voluntary but rigorous blueprint for issuers of green debt, standardizing both the definition of eligible projects and the content of allocation and impact reporting (European Commission, 2023b).

Together these measures exemplify a rules-based approach in which detailed mandates and harmonized definitions are seen as essential to ensuring that capital allocation aligns with Europe's broader societal and environmental goals. By configuring the Sustainable Finance Operating System in this way, the EU not only redirects financial flows, but also builds in accountability and comparability at every level. In addition, given the lengthy and contested process, it also more widely accepted and less likely to be substantially changed (Ahlström & Monciardini, 2022).

The United States: A Norms-Based, Market-Led Approach

In contrast, the United States has built a sustainable finance landscape driven largely by market forces, voluntary standards and private-sector innovation rather than by detailed federal mandates. The Rule-Setting Layer we will discuss in more detail in chapter 1.8 relies on principles-based guidance and market norms instead of prescriptive rules, while the Language Layer is shaped by industry-led frameworks such as the Sustainability Accounting Standards Board and the Task Force on Climate-related Financial Disclosures (Chiu, 2023). These norms define what information is material and how it should be reported, but participation is voluntary and enforcement is limited. It has less of a steering character than the framework implemented in the European Union.

At the federal level, regulators have at times proposed mandatory disclosure requirements. The rulemaking of the Securities and Exchange Commission (SEC) aimed at climate risk and greenhouse gas reporting (Cong et al., 2020). Guidance of the Department of Labor permits retirement plan fiduciaries to incorporate ESG considerations (Schanzenbach & Sitkoff, 2020). Yet these efforts have repeatedly encountered legal challenges and alternating shifts in administrative priorities.

This norms-based, market-led system affords issuers flexibility and encourages innovation in financial products and reporting approaches, but it also generates fragmentation and uneven comparability across issuers and jurisdictions. Where the European model embeds prescriptive taxonomy alignment and uniform disclosure protocols into the Operating System's core layers, the US model configures those same layers around adaptive market practices and evolving policy signals.

Core Differences: Precaution, Materiality, and Standardization

We can thus observe two fundamentally different regulatory philosophies. The European approach is rooted in the precautionary principle and the quest for standardization, favoring detailed prescriptive requirements that are established in advance. This method is designed not only to facilitate comparability and prevent greenwashing but also to ensure that financial activities are directly aligned with broader societal objectives.

By contrast, the US model embraces a high-level principles-based method that prizes flexibility and focuses primarily on disclosure and transparency regarding financial materiality. It relies on market forces and voluntary frameworks rather than mandating specific outcomes.

This philosophical divide is perhaps most evident in materiality. The EU requires double materiality reporting to capture both financial risks and environmental and social impacts. Conversely, the US focuses on financial materiality and demands disclosure only when sustainability issues affect the bottom line. These contrasting approaches influence corporate reporting practices, rating methodologies and the strategies employed by investors and regulators to integrate sustainability into financial decision-making.

Fragmentation Risk, the Brussels Effect, and International Standard-Setting

The divergence between European Union and United States frameworks creates serious compatibility issues within the Sustainable Finance Operating System. Global institutions and companies must navigate a maze of differing definitions, disclosure requirements, and regulatory expectations.

This might imply the application of double materiality reporting for EU operations while adhering to a purely financial materiality lens for US investors. It can also be observed that financial institutions use a distinctly different language for the US and EU market.

The EU's first mover work has generated a Brussels Effect, prompting many non-EU jurisdictions to adopt European benchmarks (Bradford, 2020). Yet the sheer scale and influence of US capital markets ensure that American norms continue to shape practice globally, reinforcing a landscape of competing paradigms.

International standard-setting bodies, notably the International Sustainability Standards Board, aim to bridge these divides by crafting a unified global baseline. However, their success depends on reconciling the EU's comprehensive, outcomes-oriented double materiality model with the more flexible, financially driven approach favored in the US.

1.8 The Architecture of the Sustainable Finance Operating System

The OS Metaphor: An Integrated Framework for Sustainable Finance

The Sustainable Finance Operating System is a conceptual framework to describe the interconnected network of regulations, standards, data infrastructure, and financial instruments. Although layers exist for hundreds of years in the financial industry, their existence has become more visible with the implementation of layers in the decentralized finance field with permissionless applications built on top of each other (Schär, 2021). Together, these layers govern how capital is allocated in support of environmental and social objectives:

Layer 1: Regulatory framework
Layer 2: Common Language
Layer 3: Data and Information
Layer 4: Financial Instruments

This framework offers a lens through which to analyze the interactions between rules, market behavior, data management and capital flows that collectively determine how sustainability considerations become embedded in financial decision-making.

Layer 1: The Rulebook–Regulatory Frameworks

At the foundation of the Sustainable Finance Operating System lies the rule setting layer, which establishes the essential rules, boundaries, incentives and accountability mechanisms that govern all sustainable finance activities. This layer is the rule book for capital markets, defining what counts as sustainable investment, what must be disclosed and what incentives exist to redirect capital flows. It transforms high-level policy goals into concrete requirements that shape every other component of the system.

Key instruments within this layer include mandatory regulations on reporting environmental and social impacts such as the Sustainable Finance Disclosure Regulation (SFDR) which requires funds and advisers to disclose how sustainability risks are managed. For example, the requirement for asset managers under SFDR to categorize products according to their sustainability profile has directed pools of capital toward green objectives (Cremasco & Boni, 2024).

It also encompasses policy incentives like tax credits, subsidies or guarantee programs designed to encourage investment in renewable energy and other green projects. It is rarely discussed but central banks and supervisory authorities add another dimension by embedding climate risk supervision into their oversight frameworks and by exploring adjustments to monetary policy tools to reflect environmental risks. Climate stress tests force financial institutions to assess their resilience against warming scenarios and thus realign their risk management and capital allocation practices (Acharya et al., 2023).

Layer 2: The Common Language—Disclosure Standards and Taxonomies

The second layer defines shared definitions, classifications and reporting protocols which are essential for coherent communication and comparability. While financial reporting is relatively straightforward focusing on revenues, costs and profitability, environmental and social impacts are much harder to define in commonly understood ways (e.g., Busch et al., 2021a, 2021b).

Just as a common vocabulary allows people to understand each other, disclosure standards and taxonomies enable investors, regulators, and companies to speak the same language when discussing sustainability performance. Without this layer, data would be fragmented and basically meaningless, making it impossible to measure progress or allocate capital with confidence. This layer is built on robust disclosure frameworks that set out how companies should measure and report on their environmental, social, and governance performance risks and impacts.

Complementing these reporting rules are taxonomies that classify which economic activities qualify as sustainable. The EU Taxonomy provides a science-based classification system that determines which activities make a substantial contribution to climate mitigation and adaptation without causing harm in other areas (European Commission, 2020). By codifying the criteria for sustainability, taxonomies guide both issuers and investors in identifying projects and companies that meet agreed benchmarks and help prevent the misuse of terms like green or sustainable.

Underlying both standards and taxonomies are definitions of materiality that steer what information companies disclose. Whether a company focuses on the financial impact of sustainability factors or the wider environmental and social consequences of its operations the language layer embeds these choices to guide reporting. Finally, to ensure the credibility of the information ultimately disclosed, verification standards impose assurance requirements, specifying how independent third-party reviewers must validate its accuracy and completeness. This type of ESG assurance is typically performed by large professional services.

Layer 3: Data and Information

This layer serves as the information backbone of the Sustainable Finance Operating System, responsible for collecting, processing, managing, and distributing all ESG data that drives the system. It brings together the platforms and providers that gather raw disclosures, the databases and algorithms that clean and standardize those inputs, the analytical tools that extract insights and the delivery mechanisms that make the results available to investors, regulators and other decision-makers. Without a robust information layer, the rules and language defined in the previously discussed layers cannot be translated into actionable intelligence.

At this stage, data providers collect sustainability reports and enrich this material with other information such as public filings, news feeds and corporate questionnaires. These providers often fill gaps by estimating metrics such as Scope 3 emissions or water use in supply chains. Once gathered the raw data flows into central data warehouses or cloud-based platforms where it is mapped to the shared definitions set out by the language layer.

With standardized data in place, the information layer supports impact measurement tools that calculate footprints and compare performance against benchmarks such as the Sustainable Development Goals (SDGs). Advanced analytics modules based on artificial intelligence detect risk factors or instances of greenwashing and generate metrics that feed directly into ESG ratings and risk models. In general, developments in artificial intelligence such as classification, natural language processing, or image recognition have improved the efficiency of this layer significantly.

Finally, this layer ensures that all processed information is delivered to dashboards and reporting portals. Investors draw on these feeds to allocate capital, regulators use them to monitor compliance and financial instruments activate the insights through bond and loan structures.

In this way, the information layer makes the abstract rules and definitions operational. It transforms corporate disclosures into a reliable data engine that underpins the action layer of financial instruments and market mechanisms, all within the boundaries set by regulation and guided by the common language of the system.

Layer 4: Financial Instruments

This layer represents the operational output of the Sustainable Finance Operating System in the form of financial instruments. It is where the rules established by regulators, the common language of standards and taxonomies and the processed data converge to direct real capital flows and create economic incentives aligned with sustainability objectives. In this layer, the frameworks and information pipelines are translated into practical financial activities that steer investment and pricing toward desired environmental and social outcomes.

The action layer comprises two interrelated components: specific financial instruments and broader market mechanisms. Financial instruments serve as the applications of the system, channelling funds toward targeted uses or linking financing costs to performance. Market mechanisms act as the underlying economic signals that price externalities and reward sustainable behavior across the economy.

Financial instruments include green bonds, social bonds and sustainability bonds that allocate proceeds to environmental or social projects. They also encompass sustainability-linked loans and sustainability-linked bonds whose interest costs or coupon payments adjust based on an issuer's overall performance against predefined targets. Blended finance vehicles use concessional capital to mobilize private investment in high-risk or emerging markets. Impact funds target measurable social or environmental outcomes through direct equity or debt investments. Transition finance instruments provide bespoke funding for decarbonization pathways in carbon intensive sectors. Each of these tools operates within the action layer to direct capital in alignment with the regulatory framework and the shared language of reporting protocols.

Market mechanisms create economic signals that reinforce sustainable finance goals. Carbon pricing and emissions trading systems internalize the cost of greenhouse gas emissions, sending price signals to emitters. Payments for ecosystem services monetize benefits such as water purification or habitat preservation. Social impact bonds link financial returns to the achievement of social outcomes such as improved education or health metrics. These mechanisms work alongside financial instruments to ensure that sustainability considerations are reflected in market prices and risk assessments.

References

Acharya, V. V., Berner, R., Engle, R., Jung, H., Stroebel, J., Zeng, X., & Zhao, Y. (2023). Climate stress testing. *Annual Review of Financial Economics, 15*(1), 291–326. https://doi.org/10.1146/annurev-financial-110921-101555

Achleitner, A.-K., Lutz, E., Mayer, J., & Spiess-Knafl, W. (2013). Disentangling gut feeling: Assessing the integrity of social entrepreneurs. *Voluntas: International Journal of Voluntary and Nonprofit Organizations, 24*(1), 93–124.

Ahlström, H., & Monciardini, D. (2022). The Regulatory dynamics of sustainable finance: Paradoxical success and limitations of EU reforms. *Journal of Business Ethics, 177*(1), 193–212. https://doi.org/10.1007/s10551-021-04763-x

Battiston, S., Mandel, A., Monasterolo, I., Schütze, F., & Visentin, G. (2017). A climate stress-test of the financial system. *Nature Climate Change, 7*(4), 283–288. https://doi.org/10.1038/nclimate3255

Berg, A., Ostry, J. D., Tsangarides, C. G., & Yakhshilikov, Y. (2018). Redistribution, inequality, and growth: New evidence. *Journal of Economic Growth, 23*(3), 259–305. https://doi.org/10.1007/s10887-017-9150-2

Blake, H. (2023, October 16). The Great Cash-for-Carbon Hustle. *The New Yorker*. https://www.newyorker.com/magazine/2023/10/23/the-great-cash-for-carbon-hustle

Boffo, R., & Patalano, R. (2020). *ESG investing: Practices, progress and challenges*. Éditions OCDE.

Boyes, W. J., Hoffman, D. L., & Low, S. A. (1989). An econometric analysis of the bank credit scoring problem. *Journal of Econometrics, 40*(1), 3–14. https://doi.org/10.1016/0304-4076(89)90026-2

Bradford, A. (2020). *The Brussels effect: How the European Union rules the world*. Oxford University Press.

Busch, D., Ferrarini, G., & Grünewald, S. (Eds.). (2021a). *Sustainable finance in Europe: Corporate governance, financial stability and financial markets*. Springer. https://doi.org/10.1007/978-3-030-71834-3

Busch, T., Bruce-Clark, P., Derwall, J., Eccles, R., Hebb, T., Hoepner, A., Klein, C., Krueger, P., Paetzold, F., Scholtens, B., & Weber, O. (2021b). Impact investments: A call for (re)orientation. *SN Business & Economics, 1*(2), 33. https://doi.org/10.1007/s43546-020-00033-6

Chiu, H.-Y. (2023). Sustainable finance regulation—authoritative governance or market-based governance for fund management? *Journal of Financial Transformation, 57*, 48–61.

Cong, Y., Freedman, M., & Park, J. D. (2020). Mandated greenhouse gas emissions and required SEC climate change disclosures. *Journal of Cleaner Production, 247*, Article 119111. https://doi.org/10.1016/j.jclepro.2019.119111

Cremasco, C., & Boni, L. (2024). Is the European Union (EU) Sustainable Finance Disclosure Regulation (SFDR) effective in shaping sustainability objectives? An analysis of investment funds' behaviour. *Journal of Sustainable Finance & Investment, 14*(4), 1018–1036. https://doi.org/10.1080/20430795.2022.2124838

de Freitas Netto, S. V., Sobral, M. F. F., Ribeiro, A. R. B., & da Soares, G. R. L. (2020). Concepts and forms of greenwashing: A systematic review. *Environmental Sciences Europe, 32*(1), 19. https://doi.org/10.1186/s12302-020-0300-3

Deloitte. (2023). *Financing the Green Energy Transition: A US$50 trillion catch*. https://www2.deloitte.com/content/dam/Deloitte/global/Documents/deloitte-financing-the-green-energy-transition-report-2023.pdf

Dempere, J., Alamash, E., & Mattos, P. (2024). Unveiling the truth: Greenwashing in sustainable finance. *Frontiers in Sustainability, 5*, 1362051. https://doi.org/10.3389/frsus.2024.1362051

Diamond, D. W. (1984). Financial intermediation and delegated monitoring. *The Review of Economic Studies, 51*(3), 393–414. https://doi.org/10.2307/2297430

Diaz, D., & Moore, F. (2017). Quantifying the economic risks of climate change. *Nature Climate Change, 7*(11), 774–782. https://doi.org/10.1038/nclimate3411

Durán-Román, J. L., Cárdenas-García, P. J., & Pulido-Fernández, J. I. (2021). Tourists' willingness to pay to improve sustainability and experience at destination. *Journal of Destination Marketing & Management, 19*, Article 100540. https://doi.org/10.1016/j.jdmm.2020.100540

Elton, E. J., & Gruber, M. J. (1997). Modern portfolio theory, 1950 to date. *Journal of Banking & Finance, 21*(11), 1743–1759. https://doi.org/10.1016/S0378-4266(97)00048-4

European Commission. (2018). *Renewed sustainable finance strategy and implementation of the action plan on financing sustainable growth—European Commission*. https://finance.ec.europa.eu/publications/renewed-sustainable-finance-strategy-and-implementation-action-plan-financing-sustainable-growth_en

European Commission. (2019). *Sustainability-related disclosure in the financial services sector—European Commission*. https://finance.ec.europa.eu/sustainable-finance/disclosures/sustainability-related-disclosure-financial-services-sector_en

European Commission. (2020). *EU taxonomy for sustainable activities—European Commission*. https://finance.ec.europa.eu/sustainable-finance/tools-and-standards/eu-taxonomy-sustainable-activities_en

European Commission. (2022). *Corporate sustainability reporting*. https://finance.ec.europa.eu/capital-markets-union-and-financial-markets/company-reporting-and-auditing/company-reporting/corporate-sustainability-reporting_en

European Commission. (2023a). *Eurobarometer Survey: Climate change*. https://europa.eu/eurobarometer/surveys/detail/2954

European Commission. (2023b). *The European green bond standard—Supporting the transition*. https://finance.ec.europa.eu/sustainable-finance/tools-and-standards/european-green-bond-standard-supporting-transition_en

Fung, A. (2015). Putting the public back into governance: The challenges of citizen participation and its future. *Public Administration Review, 75*(4), 513–522. https://doi.org/10.1111/puar.12361

Giese, G., Lee, L.-E., Melas, D., Nagy, Z., & Nishikawa, L. (2019). Foundations of ESG investing: How ESG affects equity valuation, risk, and performance. *Journal of Portfolio Management, 45*(5), 69–83.

Gonçalves, T. C., Dias, J., & Barros, V. (2022). Sustainability performance and the cost of capital. *International Journal of Financial Studies, 10*(3), 63. https://doi.org/10.3390/ijfs10030063

Gunningham, N. (2020). A quiet revolution: Central banks, financial regulators, and climate finance. *Sustainability, 12*(22), 9596. https://doi.org/10.3390/su12229596

Hanushek, E. A., & Woessmann, L. (2012). Do better schools lead to more growth? Cognitive skills, economic outcomes, and causation. *Journal of Economic Growth, 17*(4), 267–321. https://doi.org/10.1007/s10887-012-9081-x

Hilson, C. (2024). Climate change and the politicization of ESG in the US. *Frontiers in Political Science, 6*, 1332399. https://doi.org/10.3389/fpos.2024.1332399

Hindman, M. (2018). *The Internet trap: How the digital economy builds monopolies and undermines democracy*. 1–256.

Hummel, K., & Jobst, D. (2024). An overview of corporate sustainability reporting legislation in the European Union. *Accounting in Europe, 21*(3), 320–355. https://doi.org/10.1080/17449480.2024.2312145

Intelligent Energy Europe projects TABULA and EPISCOPE. (2017). *TABULA WebTool*. https://webtool.building-typology.eu/#bm

International Energy Agency. (2021). *Net Zero by 2050—A Roadmap for the Global Energy Sector*.

International Renewable Energy Agency. (2023). *Investment needs of USD 35 trillion by 2030 for successful energy transition*. https://www.irena.org/News/pressreleases/2023/Mar/Investment-Needs-of-USD-35-trillion-by-2030-for-Successful-Energy-Transition

Kennedy, B., & Tyson, A. (2024, December 9). How Americans View Climate Change and Policies to Address the Issue. *Pew Research Center*. https://www.pewresearch.org/science/2024/12/09/how-americans-view-climate-change-and-policies-to-address-the-issue/

Kishan, S., & Brogger, T. H. (2025, February 4). Why ESG faces backlash and its future under trump 2.0. *Bloomberg.Com*. https://www.bloomberg.com/news/articles/2025-02-04/trump-and-esg-what-does-his-presidency-mean-for-the-green-backlash

Kreibich, N. (2024). Toward global net zero: The voluntary carbon market on its quest to find its place in the post-Paris climate regime. *Wires Climate Change, 15*(5), Article e892. https://doi.org/10.1002/wcc.892

Lange, E. M., & Banadaki, N. G. (2023). ESG consideration in venture capital: Drivers, strategies and barriers. *Studies in Economics and Finance, 41*(3), 724–739. https://doi.org/10.1108/SEF-06-2023-0380

Lawrence, J., Blackett, P., & Cradock-Henry, N. A. (2020). Cascading climate change impacts and implications. *Climate Risk Management, 29*, Article 100234. https://doi.org/10.1016/j.crm.2020.100234

Levine, R. (2005). Finance and growth: Theory and evidence. In P. Aghion & S. N. Durlauf (Eds.), *Handbook of economic growth* (Vol. 1, pp. 865–934). Amsterdam: Elsevier. https://doi.org/10.1016/S1574-0684(05)01012-9

Levitsky, S., & Ziblatt, D. (2018). *How Democracies Die*. Crown. https://www.penguinrandomhouse.com/books/562246/how-democracies-die-by-steven-levitsky-and-daniel-ziblatt/

Li, S., & Kallas, Z. (2021). Meta-analysis of consumers' willingness to pay for sustainable food products. *Appetite, 163*, Article 105239. https://doi.org/10.1016/j.appet.2021.105239

Li, T.-T., Wang, K., Sueyoshi, T., & Wang, D. D. (2021). ESG: Research progress and future prospects. *Sustainability, 13*(21), 11663. https://doi.org/10.3390/su132111663

Löffler, K. U., Petreski, A., & Stephan, A. (2021). Drivers of green bond issuance and new evidence on the "greenium." *Eurasian Economic Review, 11*(1), 1–24. https://doi.org/10.1007/s40822-020-00165-y

Maxwell, A. L., & Lévesque, M. (2014). Trustworthiness: A critical ingredient for entrepreneurs seeking investors. *Entrepreneurship Theory and Practice, 38*(5), 1057–1080.

McKinsey. (2022). *COP27: Financing the transition to net zero.* https://www.mckinsey.com/capabilities/sustainability/our-insights/sustainability-blog/cop27-financing-the-transition-to-net-zero

Migliorelli, M., & Dessertine, P. (Eds.). (2020). *Sustainability and financial risks: The impact of climate change, environmental degradation and social inequality on financial markets.* Springer.

Mishkin, F. S. (2007). *The economics of money, banking, and financial markets.* Pearson education.

Neff, T., & Pickard, V. (2024). Funding democracy: Public media and democratic health in 33 countries. *The International Journal of Press/Politics, 29*(3), 601–627. https://doi.org/10.1177/19401612211060255

Ogundari, K. (2014). The paradigm of agricultural efficiency and its implication on food security in Africa: What does meta-analysis reveal? *World Development, 64*, 690–702.

Otto, I. M., Reckien, D., Reyer, C. P. O., Marcus, R., Le Masson, V., Jones, L., Norton, A., & Serdeczny, O. (2017). Social vulnerability to climate change: A review of concepts and evidence. *Regional Environmental Change, 17*(6), 1651–1662. https://doi.org/10.1007/s10113-017-1105-9

Patrinos, H. A. (2024). Estimating the return to schooling using the Mincer equation. *IZA World of Labor.* https://wol.iza.org/articles/estimating-the-return-to-schooling-using-the-mincer-equation/long

Piketty, T. (2018). *Capital in the twenty-first century.* Harvard University Press.

Psaki, S. R., Chuang, E. K., Melnikas, A. J., Wilson, D. B., & Mensch, B. S. (2019). Causal effects of education on sexual and reproductive health in low and middle-income countries: A systematic review and meta-analysis. *SSM—Population Health, 8*, Article 100386. https://doi.org/10.1016/j.ssmph.2019.100386

Reardon, S. F. (2013). *The widening income achievement gap.* Structure of Schooling: Readings in the Sociology of Education.

Rennert, K., Errickson, F., Prest, B. C., Rennels, L., Newell, R. G., Pizer, W., Kingdon, C., Wingenroth, J., Cooke, R., Parthum, B., Smith, D., Cromar, K., Diaz, D., Moore, F. C., Müller, U. K., Plevin, R. J., Raftery, A. E., Ševčíková, H., Sheets, H., & Anthoff, D. (2022). Comprehensive evidence implies a higher social cost of CO_2. *Nature, 610*(7933), 687–692. https://doi.org/10.1038/s41586-022-05224-9

Rennert, K., Kingdon, C., & Prest, B. C. (2025). *Social Cost of Carbon 101.* https://www.rff.org/publications/explainers/social-cost-carbon-101/

Reuters. (2024, October 21). WisdomTree pays $4 million over SEC "greenwashing" charges. *Reuters*. https://www.reuters.com/business/finance/wisdomtree-pays-4-million-over-sec-greenwashing-charges-2024-10-21/

Santos, C., Coelho, A., & Marques, A. (2024). A systematic literature review on greenwashing and its relationship to stakeholders: State of art and future research agenda. *Management Review Quarterly, 74*(3), 1397–1421. https://doi.org/10.1007/s11301-023-00337-5

Schanzenbach, M. M., & Sitkoff, R. H. (2020). Reconciling fiduciary duty and social conscience: The law and economics of ESG investing by a trustee. *Stanford Law Review, 72*, 381.

Schär, F. (2021). Decentralized finance: On blockchain and smart contract-based financial markets. *Review of the Federal Reserve Bank of St Louis, 103*(2), 153–174.

Schoenmaker, D., & Schramade, W. (2018). *Principles of sustainable finance*. Oxford University Press.

Schoenmaker, D., & Schramade, W. (2023). *Corporate finance for long-term value*. Springer. https://doi.org/10.1007/978-3-031-35009-2

SEC. (2024a). *In the Matter of WisdomTree Asset Management, Inc., Administrative Proceeding File No. 3-22268, Investment Advisers Act Release No. 6753.*

SEC. (2024b). *SEC Charges Advisory Firm WisdomTree with Failing to Adhere to Its Own Investment Criteria For ESG-Marketed Funds*. https://www.sec.gov/newsroom/press-releases/2024-173

Söderholm, P. (2012). Modeling the economic costs of climate policy: An overview. *American Journal of Climate Change, 1*(1), 14–32.

Spiess-Knafl, W., & Scheck, B. (2023). *Impact investing: Instruments, mechanisms and actors* (2nd ed.). Palgrave Macmillan.

Starks, L. T. (2023). Presidential address: Sustainable finance and ESG issues—value versus values. *The Journal of Finance, 78*(4), 1837–1872. https://doi.org/10.1111/jofi.13255

Vladimirova, K., & Le Blanc, D. (2016). Exploring links between education and sustainable development goals through the lens of UN flagship reports. *Sustainable Development, 24*(4), 254–271. https://doi.org/10.1002/sd.1626

Wang, Y., Delgado, M. S., & Xu, J. (2023). When and where does it pay to be green? A look into socially responsible investing and the cost of equity capital. *International Journal of Corporate Social Responsibility, 8*(1), 1. https://doi.org/10.1186/s40991-023-00077-5

Zaccone, M. C., & Pedrini, M. (2020). ESG factor integration into private equity. *Sustainability, 12*(14), 5725. https://doi.org/10.3390/su12145725

CHAPTER 2

Foundations of Sustainable Finance: Theory and Practice

Abstract This chapter explores the theoretical foundations and practical applications underpinning sustainable finance. It begins by examining the historical evolution and ongoing debate surrounding corporate purpose, contrasting traditional shareholder primacy with broader stakeholder perspectives and exploring different theoretical arguments. The chapter traces the evolving definition of sustainable finance, mapping its journey from early ethical exclusions ("sin stocks") to the comprehensive integration of Environmental, Social and Governance (ESG) factors and the targeted approach of impact investing aligned with goals like the Sustainable Development Goals (SDGs). Building on these foundations, the chapter analyzes the mechanisms through which various stakeholders exert influence to drive corporate change toward sustainability. The chapter then explores the impacts of sustainable finance on capital markets, detailing how ESG considerations are reshaping equity valuations, debt instruments (like green and social bonds) as well as the overall cost of capital for companies. Finally, the chapter provides a snapshot of the sustainable finance market, outlining its current size, growth trends, dominant players, evolving fee structures and key asset allocation patterns across equities, fixed income, and alternatives.

Keywords Corporate purpose · Evolution of sustainable finance · Market size

W. Spiess-Knafl, *The Sustainable Finance Operating System*, Palgrave Studies in Impact Finance,
https://doi.org/10.1007/978-3-031-97499-1_2

2.1 Defining Corporate Purpose: Shareholder vs. Stakeholder Views

Historical Evolution: From Multiple Constituencies to Shareholder Primacy and Back

At the core of sustainable finance is the question about the appropriate objective function or the ultimate goal of business. Historically, three competing approaches can be observed: (1) shareholder primacy, which prioritizes investor returns, (2) management discretion, which trusts professional managers to balance complex decisions and (3) social responsibility, which considers obligations to employees and the broader public. (Bratton & Wachter, 2008).

The evolution of corporate purpose has undergone significant shifts over time. Prior to the 1970s, corporations were broadly understood to serve multiple constituencies. The landmark case A.P. Smith Manufacturing Co. v. Barlow (1953) established that corporations could make charitable contributions even when these were not directly benefiting shareholders. That means that corporate philanthropy was legally sanctioned (Pearce, 2015).

However, as economic theory developed in the 1970s and 1980s, agency theory and the efficient market hypothesis increasingly focused corporate governance on shareholder returns (Fama, 1970; Jensen & Meckling, 1976). It was most famously advocated by Milton Friedman arguing that a company's primary duty is to maximize profits or market value (Friedman, 1970). This culminated in the shareholder value movement of the 1990s (Fligstein & Shin, 2007). This tendency was best exemplified by Jack Welch's leadership at GE, before witnessing a gradual return to broader perspectives in the 2000s and 2010s.

This shift to broader perspectives can also be seen in the context of finance. More recent scholarship, however, challenges this narrow view focused on shareholder value by arguing that a broader focus on shareholder welfare is more appropriate (Hart & Zingales, 2017). This perspective recognizes that shareholders often care not only about financial returns but also about ethical, social and environmental issues.

The shareholder-focused school of thought rests on the assumption that profit-making activities and ethical or social concerns can be neatly separated. However, Hart and Zingales (2017) argue that shareholders internalize externalities in their personal lives such as choosing to drive an electric car or purchasing fair trade coffee. Therefore, shareholders might prefer companies to adopt similar ethical and socially responsible behaviors.

Why Shareholders Might Care About More than Profit

Economic theory offers several explanations for why even purely rational shareholders might endorse socially responsible corporate actions. Elhauge (2005) shows that shareholders can legitimately vote for policies that reduce short-term profits if those same policies avert larger social harms, effectively trading modest financial sacrifice for greater collective welfare. Zivin and Small (2005) extend this logic by treating corporate philanthropy as a form of pooled giving. Firms can donate more efficiently than individuals thanks to economies of scale or tax advantages. In this way, shareholders benefit from the higher overall impact per dollar contributed. They call responsible firms "charity-investment bundle."

Morgan and Tumlinson (2019) show that shareholders who value public goods can direct firms to sacrifice profits. The idea is that firms can more effectively provide these goods than shareholders by acting individually by centralizing contributions and abating production harms. This improves overall shareholder welfare, even though direct financial returns are lower.

Another aspect is formalized by Opp and Oehmke (2020). They show how investors who care about externalities can steer capital toward companies that internalize social and environmental costs, using their allocation choices to reward responsible behavior and penalize neglect.

Purpose in Practice: The Boardroom Battleground

The boardroom has become the battleground where competing theories of corporate purpose are tested and ultimately decided by courts. The tension between maximizing short-term market value and securing long-term shareholder welfare was especially apparent in recent years when companies were forced to cut costs. A good example is the widespread rolling back of programs focused on diversity, equity, and inclusion (DEI) at the beginning of the second Trump presidency (Davis, 2025).

Global governance frameworks are evolving to reconcile profit with purpose. In the European Union, the Corporate Sustainability Reporting

Directive (CSRD) and the Corporate Sustainability Due Diligence Directive (CSDDD) in its first version, require directors to factor in ESG risks and even oversee supply chain impacts. In the United States, Benefit Corporation statutes and B Corp certifications legally protect leaders who balance public benefits with profits.

Investor voices and market signals also drive change. One example is the annual letters sent by the CEO of BlackRock. In his 2020 CEO letter, he emphasized that "climate risk is investment risk" and warned that companies must have a clear social purpose or risk losing their license to operate (Fink, 2020).

Critiques and Counterarguments: Accountability and Legitimacy Debates

While shareholder welfare approaches gain traction, they face significant challenges. The seminal stakeholder model introduced by Freeman (1984) argues that managers owe duties not only to shareholders but also to employees, customers, suppliers and communities. Yet critics such as Friedman (1970) and Bebchuk and Tallarita (2020) warn that this expansion of corporate purpose risks turning executives into unelected policymakers without democratic legitimacy. For example, how should managers balance corporate climate commitments with shareholder interests in regions where tax income depends on the sale of oil and gas.

From an accountability standpoint, Jensen (2001) argues that stakeholder mandates fragment a firm's focus and create agency costs. This lack of a single, clear objective function for balancing stakeholder interests ultimately renders managers unaccountable, empowering them to potentially use firm resources according to their own preferences.

Despite these critiques, proponents offer compelling counterarguments: they point to pervasive externalities and coordination failures that markets alone cannot resolve (Tirole, 2017). They also champion enhanced shareholder voting mechanisms to legitimize sustainability decisions (Hart & Zingales, 2017). There are also frameworks which see purpose and profits are complementary rather than conflicting (Edmans, 2020).

2.2 Sustainable Finance: An Evolving Definition

Sustainable finance refers to the process of incorporating environmental, social and governance (ESG) factors into financial decision-making with the dual aim of achieving long-term financial performance and contributing to sustainable development (e.g., Ziolo et al., 2019).

The evolution of sustainable finance reflects the gradual broadening of focus from purely moral or ethical concerns to a more comprehensive integration of environmental, social and governance factors in investment decisions (Cunha et al., 2021). It is also interesting to note that even within single decades there are substantial changes to areas of interest (Alshater et al., 2023).

During the seventeenth to nineteenth centuries, early forms of socially responsible investing emerged from religious and moral imperatives. Religious groups such as the Quakers and Methodists pioneered this approach by refusing to invest in activities that conflicted with their ethical beliefs, such as the slave trade or the production of weapons. These early investors relied on negative screening methods, deliberately avoiding industries they deemed "sinful" or harmful (Monahan, 2002).

In the period before 1970, there was no specific term for what we now call sustainable finance. During this time, investors that often were religious organizations or churches imposed restrictions on investments that conflicted with their ethical codes, commonly avoiding so-called "sin stocks" such as those in the alcohol, tobacco, and gambling sectors (Schueth, 2003).

The 1970s saw the emergence of the term "socially responsible investment." Driven by the social and political movements of the previous decade, including opposition to militarism and racism, investors began to focus on supporting companies engaged in corporate social responsibility and divesting from firms that benefited from controversial practices, such as those profiting from the Vietnam War or apartheid in South Africa (Teoh et al., 1999). Many organizations that still exist today have their roots in this time.

The period from the 1980s to the 1990s witnessed a significant broadening of ESG awareness. High-profile environmental disasters, such as the 1984 Bhopal gas leak and the 1989 Exxon Valdez oil spill, raised public and investor consciousness about the importance of corporate environmental responsibility. In 1987, the Brundtland Commission's report titled "Our Common Future" introduced the concept of sustainable development by linking environmental protection with social equity and economic growth. This was further reinforced by the 1992 Rio Earth Summit, which helped shape the global agenda on sustainable development and encouraged businesses and investors to acknowledge environmental constraints and social responsibilities (Pollman, 2024).

The field diversified further. The term "social investment" appeared as the first funds were labeled to address broader societal concerns, including human rights, gender equality and sexual diversity. At the same time, "ethical investment" became popular as corporations grew more interested in business ethics. Ethical investment largely relied on negative screening methods avoiding industries or companies based on moral judgments (Cunha et al., 2021).

In the 1990s, the focus shifted toward the environmental dimension, with the terms "green investment" and "green finance" entering the lexicon. These concepts emphasized promoting positive environmental impacts through investment strategies and laid the groundwork for later developments in climate finance (Cunha et al., 2021). It also shows that decades are needed to implement policy changes.

The 2000s marked the institutionalization of sustainable finance through the development of global frameworks and UN-backed initiatives. The launch of the Principles for Responsible Investment (PRI) in 2006 provided a voluntary framework for integrating ESG factors into investment analysis, while the Equator Principles, introduced in 2003, offered a risk-management framework for assessing and managing social and environmental risks in project finance.

It was also the time when "impact investments" were first mentioned (e.g., Höchstädter & Scheck, 2015). During this period, the number of ESG-focused funds grew significantly, as investors expanded their criteria beyond negative screening to include shareholder engagement, best-in-class strategies and thematic investing. This widening scope reflected a more holistic approach to integrating sustainability into the investment process.

In the 2010s, the terminology further refined itself in response to the urgent challenges of climate change. "Climate finance" emerged to describe efforts aimed at both mitigating and adapting to climate change, while "low-carbon investment" or "low-carbon finance" specifically focused on supporting companies that operate with lower carbon emissions or on divesting from high-emitting industries (Diaz-Rainey et al., 2017; Zhang et al., 2019; Campiglio, 2016; Geddes et al., 2018).

Additionally, "blue finance" was coined to address the financing needs of ocean conservation initiatives (Wabnitz & Blasiak, 2019). Finally, the term "ESG investment" encapsulates the broader approach, combining environmental and social factors with a critical component of corporate governance, which ensures that socio-environmental concerns are adequately addressed in investment strategies (Busch et al., 2016; Daugaard, 2020).

The adoption of the Paris Agreement in 2015, alongside the United Nations Sustainable Development Goals (SDGs), intensified the focus on climate change, renewable energy and social impact among investors. The green bond market, which had been initiated by development banks in the late 2000s, experienced strong growth as both corporations and governments increasingly turned to green bonds to finance climate-friendly projects. The Task Force on Climate-related Financial Disclosures (TCFD) which was established in 2015 and disbanded in 2023 further pushed companies and investors toward greater transparency regarding climate risks and opportunities. Mainstream asset managers began integrating ESG metrics into their investment strategies, driven by mounting evidence that sustainable business practices can enhance long-term financial performance (e.g., Friede et al., 2015).

From the 2020s to the present, sustainable finance has continued to evolve amid significant regulatory momentum and accelerated market demand. In Europe, the EU Action Plan on Sustainable Finance aims to reorient capital flows toward sustainable investments, introducing measures such as a taxonomy for sustainable economic activities. Concurrently, global initiatives by the IFRS Foundation and the International Sustainability Standards Board (ISSB) are working toward standardized sustainability reporting frameworks.

Institutional and retail investors have shown heightened interest in sustainable investment products, resulting in record inflows into ESG funds. However, this growth has also brought challenges, such as increased scrutiny over greenwashing and persistent issues with data standardization.

2.3 Driving Change: How Stakeholders Exert Influence

There are different strategies for how stakeholders can influence companies. Responsible ownership and direct shareholder engagement is one approach to influence corporate strategies. Shareholders leverage their equity stake to initiate corporate change. Through detailed ESG-focused analysis and ongoing dialogue they can try to shape board composition or tie executive compensation to sustainability targets. Obviously, larger shareholders will find it easier to engage the management of the company. This active stewardship and engagement can lead to measurable financial performance as firms respond to the prospect of binding votes against directors or major proposals (Dimson et al., 2015).

Smaller shareholders can file shareholder resolutions such as those demanding specific climate risk disclosures or board diversity targets, collaborating with other investors. These strategies can often be very successful and receive widespread media attention (Ambrose, 2024; Conlon, 2024).

Stakeholders such as NGOs or consumers can also influence corporate decisions. For example Target's rollback of diversity, equity and inclusion initiatives led to sustained declines in foot traffic and sales as consumers boycotted stores (Bacon, 2025).

Financial activism harnesses use market-based tools to reward or penalize corporate behavior. Divesting or underweighting companies with weak ESG records sends a clear signal and can potentially raise those firms' cost of capital due to perceived higher risk or reduced investor demand. Conversely, strategies like positive screening, selecting ESG leaders and overweighting their shares create incentives that encourage better performance. Collaborative initiatives such as Climate Action 100 plus amplify these effects by coordinating engagement with the world's highest emitting companies and tracking their progress toward net zero targets.

Legal and regulatory pressure leverages litigation and rulemaking to enforce change. Climate-related lawsuits against governments and companies that fail to disclose risk or prevent environmental harm have put firms on notice that failure to adapt can lead to steep legal and reputational costs. This risk incentivizes not only corporations but also financial institutions to adjust their lending and investment practices.

2.4 Market Impacts: How Sustainable Finance Reshapes Capital and Companies

The stakeholder influence mechanisms outlined above generate tangible consequences within capital markets and shape corporate behavior. Growing awareness of environmental and social risks as well as investor and regulatory pressure is reshaping equity as well as debt markets. There are different pathways through which sustainable finance influences valuations, cost of capital and fund flows within capital markets.

On the equity capital markets, heightened investor demand for companies with strong ESG performance translates into increased share prices with all related benefits.

This is driven by investors with a “values” as well as “value” lens. Investors who want to align their portfolios with their values increase the demand for companies in certain areas such as renewable energy or education or for those companies with high ESG scores. Investors with a value lens build their portfolio following a belief that sustainability correlates with lower risks and potential competitive advantage.

In this sense, improved transparency driven by evolving disclosure standards and sophisticated ESG rating systems enhances market efficiency. In addition, better comparability allows investors to distinguish between genuine sustainability leaders and laggards, leading to more informed capital allocation decisions and more accurate pricing of long-term risks.

Evidence indicates that firms with superior ESG credentials often enjoy a lower cost of equity capital for various reasons (e.g., Wang et al., 2023). This financing advantage contributes to higher valuations, reinforcing the cycle as a higher valuation might open new opportunities as discussed in chapter **Fehler! Verweisquelle konnte nicht gefunden werden.**

On the debt capital markets side, new instruments such as green, social and sustainability-linked instruments have emerged as a major channel for directing fixed income capital toward environmental and social objectives. These bonds enable issuers to raise funds explicitly for projects such as renewable energy, affordable housing and water infrastructure.

Strong investor demand for sustainable bonds often results in slightly lower yields compared to conventional debt, known as a greenium, reflecting investor willingness to accept marginally reduced returns in exchange for positive impact. For example, in the first of 2023, 32% of green bonds achieved a greenium (Climate Bonds Initiative, 2023).

The greenium creates a strong financial incentive for corporations and governments to issue sustainable bonds. Lower financing costs directly benefit issuers and help expand the market for projects that address climate change, social inequality and other global challenges. There is demand for these bonds as evidenced in the market.

The NextGenerationEU green bond which was issued by the European Commission in 2021 and raised €12 billion for green and sustainable investments was 11 times oversubscribed (European Commission, 2021). German development bank reported that a 10-year green bond with a volume of €3 billion was 10 times oversubscribed (KfW, 2023). For the United States, it was found that green bonds are on average 5 times oversubscribed, while vanilla equivalents are oversubscribed 3 times (Harrison, 2023).

Standardized guidelines such as the Green Bond Principles and the EU Green Bond Standard together with enhanced regulatory oversight bolster investor confidence and market integrity. These frameworks ensure consistency in the use of proceeds and impact reporting, facilitating growth in sustainable bond issuance. As investors increasingly channel funds into sustainable bonds, it also means that the reallocation often increases demand for these assets and supports a wider transition toward sustainable economic activities.

2.5 The Sustainable Finance Market: Size, Players, and Trends

Measuring the Market: Size, Growth, and Methodological Challenges

The global sustainable investment market reached approximately 30.3 trillion USD in assets under management (AUM) at the start of 2022, according to the Global Sustainable Investment Alliance's (2023) aggregation of regional surveys, manager submissions and public disclosures. However, interpreting this headline figure demands care: in the United States alone, AUM fell from 17.1 trillion USD in 2020 to 8.4 trillion USD in 2022 after US SIF tightened its methodology to exclude funds lacking explicit ESG criteria driven by greenwashing scrutiny (Cortese, 2022).

When the US is set aside, the volume of sustainable assets under management across Europe, Canada, Japan, Australia and New Zealand grew by 20 percent between 2020 and 2022, climbing from 18.2 trillion USD to 21.9 trillion USD. This underlying expansion reflects both "value"-led investments where ESG integration mitigates financial risks and "values"-driven allocations focused on social or environmental impact (Global Sustainable Investment Alliance, 2023).

Regionally, Europe's sustainable AUM increased from 12 trillion USD to 14.1 trillion USD, yet the share of total managed assets edged down slightly to 38 percent, as stricter labeling rules such as the EU's Sustainable Finance Disclosure Regulation (SFDR) and a maturing market definition led to a reclassification of assets.

The report also states that Canada has seen a drop of the sustainable investing assets from 62 to 47 percent. In contrast, Japan's sustainable share rose to 34 percent (with AUM moving from 2.9 trillion USD to 4.3 trillion USD) and Australia/New Zealand climbed to 21 percent. Elsewhere in Asia, nascent markets like China and Taiwan are beginning to contribute meaningful new capital, though their AUM remains measured in the tens of billions and mostly consist of green loans from commercial bank.

Although methodological revisions complicate straight-line comparisons, the data make clear that sustainable finance continues to build a substantial, growing asset base worldwide, even as definitions, standards and data practices evolve.

Dominant Players

A handful of firms dominate sustainable fund distribution as shown in Table 2.1. These large managers shape product availability, fee structures and investor choice. Their scale drives down average fees through competitive pressure. The growth of passive ESG ETFs benefits "value"-seeking investors but can potentially crowd out smaller innovators catering to "values"-driven clients.

European managers enjoy an early regulatory head start, while US firms still dominate due to their large global footprint. It is also interesting to note that asset managers use distinctly different languages in the US and the EU due to public perceptions.

However, there are also many small asset managers in the private equity and debt markets which are managing funds in the range of 10–50 million dollar or euro (Spiess-Knafl & Scheck, 2019). These funds are managing early-stage investments in impact-oriented enterprises.

Table 2.1 Overview of asset managers with activities in sustainable finance

Asset manager	*Total AUM*	*Sustainable/responsible/ESG AUM*	*Key sustainable commitments/strategy highlights*
BlackRock	~ $10 Trillion +	$705 Billion	World's largest manager; provides client choice via sustainable platform (Screened, Uplift, Thematic, Impact); ESG integration; stewardship; Net Zero commitment; faces ESG backlash
Amundi	€2.2 Trillion	€983 Billion ($1 T +)	Largest European manager; 100% ESG integration target for active funds; Ambitions 2025 plan; Net Zero commitment; strong ESG voting record
Vanguard	~ $8–9 Trillion +	Not disclosed	Second largest global manager; primarily passive focus; offers some exclusionary funds; influential stewardship due to scale; lower ESG voting support vs European peers
Fidelity Investments	~ $12 Trillion +	~ $4.2 Billion; Broader ESG considerations likely much larger	Major global manager; offers dedicated sustainable funds; increasing focus on ESG data/research; emphasizes long-term value; reports ESG outperformance potential
State Street Global Advisors	~ $4.1 Trillion +	Not disclosed	Major index/ETF provider; offers ESG strategies; asset stewardship focus; TCFD/SASB/GRI reporting; provides proxy-voting choice

(continued)

Table 2.1 (continued)

Asset manager	*Total AUM*	*Sustainable/responsible/ESG AUM*	*Key sustainable commitments/strategy highlights*
Capital Group	~ $2.5 Trillion +	Not disclosed	Large active manager; reports high ESG adoption among institutional clients; integrates ESG for risk/opportunity assessment
Morgan Stanley	~ $1.5 Trillion +	Significant focus via Calvert subsidiary, amount not disclosed	Calvert is a leader in responsible investing; MS Investment Management integrates ESG; emphasizes financial materiality of sustainability; offers thematic/impact options
JPMorgan Asset Management	~ $3 Trillion +	Not disclosed	Major global manager; active in green finance underwriting integrates ESG considerations; offers sustainable products
Goldman Sachs Asset Management	~ $2.8 Trillion +	Significant clean energy/climate commitments, $325 billion in sustainability-related assets under supervision	Active in sustainable/impact investing; integrates ESG; provides thematic research
Federated Hermes	~ $750 Billion +	$50.8 Billion (Managed by Federated Hermes Ltd, incl. Hermes GPE)	Strong historical focus on stewardship/engagement via Hermes EOS; offers dedicated sustainable funds (e.g., climate credit); integrates ESG

Sources BlackRock (2025), Amundi (2025), Fidelity Investments (2025), Vanguard (2025), Capital Group (2025), Morgan Stanley (2025), Morgan (2025), Goldman Sachs (2025), FederatedHermes (2025)

Management Fees

Investors in sustainable funds still pay a greenium relative to conventional peers. At the end of 2023, the asset-weighted average expense ratio for sustainable funds was 0.52 percent versus 0.36 percent for traditional funds. That is a gap of 16 basis points. Yet these fees have been falling sharply: over the past decade, the average fee charged by sustainable funds has declined by roughly 35 percent and the average fee actually paid by investors has dropped about 44 percent, reflecting both equal- and asset-weighted compression (Morningstar, 2024).

However, the trend is less pronounced among funds with the most demanding sustainability mandates such as Article 9 products under SFDR which incur higher costs for impact verification, specialized research and active engagement. Overall, the broad decline in fees makes sustainable strategies more affordable for a wider range of investors while placing renewed margin pressure on active managers across the ESG spectrum.

Again, in private equity markets it is still not uncommon to have management fees of around 2% as fund volumes are smaller and the investment process generally involves more work.

Asset Allocation Trends

Precise allocation figures are hard to pin down as surveys cover varying universes such as asset owners vs. retail funds and regional scopes. Definitions continue to evolve as outlined above.

Equities remain a cornerstone of sustainable portfolios, often representing the largest part of portfolio allocation. This dominance reflects the long history of ESG integration in public markets, the high availability of corporate sustainability data and a variety of equity-focused ESG products. Their high liquidity and transparent pricing make equities the natural entry point for both "value" investors seeking risk mitigation and "values" investors pursuing thematic impact.

Fixed income allocations are a smaller proportion than equity, although one global survey of institutional asset owners report allocations as high as 53 percent to labeled sustainable debt (FTSE Russell, 2022). While the precise share is hard to establish, the growth trajectory of sustainable fixed income is undeniable. Fixed income products such as green bonds, social bonds, sustainability-linked and transition instruments now cater to a broader spectrum of risk-return profiles and rest on increasingly rigorous issuance and verification standards.

Annual issuance in 2024 reached 1.1 trillion USD bringing cumulative labeled sustainable debt to 6.2 trillion USD by December 2024, with green bonds alone representing 57 percent of that total (World Bank,

2025). Compare that number to the total 30.3 trillion USD in assets under management established by the Global Sustainable Investment Alliance's (2023).

Alternatives such as real assets, private equity and venture capital make up the remainder. Their smaller share reflects the inherently lower liquidity of private markets and ongoing challenges in ESG data availability and standardization for unlisted assets. Even so, dedicated climate tech, renewable infrastructure and impact-investing funds are among the fastest-growing segments, attracting both "value"-driven investors seeking long-term growth themes and "values"-driven investors who are attracted to more direct pathways.

References

Alshater, M. M., Atayah, O. F., & Hamdan, A. (2023). Journal of sustainable finance and investment: A bibliometric analysis. *Journal of Sustainable Finance & Investment, 13*(3), 1131–1152. https://doi.org/10.1080/20430795.2021.1947116

Ambrose, J. (2024, January 16). Shell faces shareholder rebellion over climate activist resolution. *The Guardian.* https://www.theguardian.com/business/2024/jan/16/shell-faces-shareholder-rebellion-over-climate-activist-resolution

Amundi. (2025). *2024 Integrated Report: Shaping the future together, with savings solutions tailored to your needs.* https://about.amundi.com/files/nuxeo/dl/83617b1c-0685-4d3e-87b9-9ed10587d70a

Bacon, A. (2025, April 21). *Target rolled back DEI efforts. A boycott ensued—and traffic dropped.* CNN. https://www.cnn.com/2025/04/21/business/target-boycott-jamal-bryant-minority-businesses

Bebchuk, L. A., & Tallarita, R. (2020). The illusory promise of stakeholder governance. *Cornell Law Review, 106*, 91.

BlackRock. (2025). *Sustainable and Transition Investing.* BlackRock. https://www.blackrock.com/au/solutions/sustainable-transition-investing

Bratton, W. W., & Wachter, M. L. (2008). Shareholder primacy's corporatist origins: Adolf Berle and the modern corporation. *Journal of Corporation Law, 34*, 99.

Busch, T., Bauer, R., & Orlitzky, M. (2016). Sustainable Development and Financial Markets: Old Paths and New Avenues. *Business & Society, 55*(3), 303–329. https://doi.org/10.1177/0007650315570701

Campiglio, E. (2016). Beyond carbon pricing: The role of banking and monetary policy in financing the transition to a low-carbon economy. *Ecological Economics, 121*, 220–230.

Capital Group. (2025). *Environmental, Social and Governance (ESG)*. Capital Group EACG. https://www.capitalgroup.com/institutions/at/en/investments/esg.html

Climate Bonds Initiative. (2023, September 19). *32% of Green Bonds Achieved a Greenium in the First Half of 2023*. Climate Bonds Initiative. https://www.climatebonds.net/resources/press-releases/2023/09/32-green-bonds-achieved-greenium-first-half-2023

Conlon, R. (2024, October 23). Why these nuns have filed more than 350 shareholder resolutions. *NPR*. https://www.npr.org/2024/10/18/nx-s1-5150123/why-these-nuns-have-filed-more-than-350-shareholder-resolutions

Cortese, A. (2022, December 14). *The case of the suddenly missing trillions in ESG investments*. ImpactAlpha. https://impactalpha.com/the-case-of-the-suddenly-missing-trillions-in-esg-investments/

Cunha, F. A. F. de S., Meira, E., & Orsato, R. J. (2021). Sustainable finance and investment: Review and research agenda. *Business Strategy and the Environment*, *30*(8), 3821–3838. https://doi.org/10.1002/bse.2842

Daugaard, D. (2020). Emerging new themes in environmental, social and governance investing: A systematic literature review. *Accounting & Finance, 60*(2), 1501–1530. https://doi.org/10.1111/acfi.12479

Davis, D.-M. (2025, April 17). Here are all the tech companies rolling back DEI or still committed to it—So far. *TechCrunch*. https://techcrunch.com/2025/04/17/here-are-all-the-tech-companies-rolling-back-dei-or-still-committed-to-it-so-far/

Diaz-Rainey, I., Robertson, B., & Wilson, C. (2017). Stranded research? Leading finance journals are silent on climate change. *Climatic Change, 143*(1–2), 243–260. https://doi.org/10.1007/s10584-017-1985-1

Dimson, E., Karakaş, O., & Li, X. (2015). Active ownership. *The Review of Financial Studies, 28*(12), 3225–3268. https://doi.org/10.1093/rfs/hhv044

Edmans, A. (2020). *Grow the Pie: How great companies deliver both purpose and profit*. Cambridge University Press. https://doi.org/10.1017/9781108860093

Elhauge, E. (2005). Sacrificing corporate profits in the public interest. *New York University Law Review, 80*, 733.

European Commission. (2021). *NextGenerationEU first green bond issuance*. European Commission - European Commission. https://ec.europa.eu/commission/presscorner/detail/pt/qanda_21_5211

Fama, E. F. (1970). Efficient capital markets. *Journal of Finance, 25*(2), 383–417.

Federated Hermes. (2025). *About us*. US. https://www.hermes-investment.com/us/en/professional/about-us/

Fidelity Investments. (2025). *Sustainable Investing | ESG*. https://www.fidelity.com/mutual-funds/sustainable

Fink, L. (2020). *Larry Fink's Letter to CEOs*. BlackRock. https://www.blackrock.com/corporate/investor-relations/2020-larry-fink-ceo-letter

Fligstein, N., & Shin, T. (2007). Shareholder value and the transformation of the U.S. economy, 1984–2000. *Sociological Forum, 22*(4), 399–424. https://doi.org/10.1111/j.1573-7861.2007.00044.x

Freeman, R. E. (1984). *Strategic management: A stakeholder approach*. Pitman Publishing Inc.

Friede, G., Busch, T., & Bassen, A. (2015). ESG and financial performance: Aggregated evidence from more than 2000 empirical studies. *Journal of Sustainable Finance & Investment, 5*(4), 210–233.

Friedman, M. (1970). The Social Responsibility of Business is to Increase its Profits. *The New York Times Magazine*.

FTSE Russell. (2022). *2022 global survey findings from asset owners*. https://www.lseg.com/content/dam/ftse-russell/en_us/documents/reports/global-asset-owner/sustainable-investment-2022-global-survey-findings-from-asset-owners.pdf

Geddes, A., Schmidt, T. S., & Steffen, B. (2018). The multiple roles of state investment banks in low-carbon energy finance: An analysis of Australia, the UK and Germany. *Energy Policy, 115*, 158–170.

Global Sustainable Investment Alliance. (2023). *Global Sustainable Investment Review 2022*. https://www.gsi-alliance.org/members-resources/gsir2022/

Goldman Sachs. (2025). *Climate Transition and Inclusive Growth*. Goldman Sachs Asset Management. https://am.gs.com/en-us/institutions/creating-impact

Harrison, C. (2023). *Green bond pricing in the primary market H2022*. Climate Bonds Initiative.

Hart, O., & Zingales, L. (2017). Companies should maximize shareholder welfare not market value. *Social Science Research Network*. https://doi.org/10.2139/ssrn.3004794

Höchstädter, A. K., & Scheck, B. (2015). What's in a name: An analysis of impact investing understandings by academics and practitioners. *Journal of Business Ethics, 132*(2), 449–475.

Jensen, M. (2001). Value maximisation, stakeholder theory, and the Corporate Objective Function. *European Financial Management, 7*(3), 297–317. https://doi.org/10.1111/1468-036X.00158

Jensen, M. C., & Meckling, W. H. (1976). Theory of the firm: Managerial behavior, agency costs and ownership structure. *Journal of Financial Economics, 3*(4), 305–360.

J.P. Morgan. (2025). *ESG*. https://www.jpmorgan.com/investment-banking/esg-at-j-p-morgan

KfW. (2023). *Transparency in "Green Bonds—Made by KfW."* https://www.kfw.de/About-KfW/Newsroom/Latest-News/Pressemitteilungen-Details_758464.html

Monahan, M. (2002). The ethics of socially responsible investing. *Business & Professional Ethics Journal, 21*(3/4), 27–46.

Morgan, J., & Tumlinson, J. (2019). Corporate provision of public goods. *Management Science, 65*(10), 4489–4504. https://doi.org/10.1287/mnsc.2018.3137

Morgan Stanley. (2025). *Institutional Investor.* Morgan Stanley Investment Management. https://www.morganstanley.com/im/en-us/institutional-investor.html

Morningstar. (2024). *Morningstar's 2023 Annual US Fund Fee Study.* Morningstar, Inc. https://www.morningstar.com/lp/annual-us-fund-fee-study

Opp, M., & Oehmke, M. (2020). A theory of socially responsible investment. *Centre for Economic Policy Research Discussion Paper, 14351.* https://repec.cepr.org/repec/cpr/ceprdp/DP14351.pdf

Pearce, J. A. (2015). The rights of shareholders in authorizing corporate philanthropy. *Villanova Law Review, 60*, 251.

Pollman, E. (2024). The making and meaning of ESG. *Harvard Business Law Review, 14*, 403.

Schueth, S. (2003). Socially responsible investing in the United States. *Journal of Business Ethics, 43*(3), 189–194. https://doi.org/10.1023/A:1022981828869

Spiess-Knafl, W., & Scheck, B. (2019). *Social enterprise finance market analysis and recommendations for delivery options.* Publications Office of the European Union.

Teoh, S. H., Welch, I., & Wazzan, C. P. (1999). The effect of socially activist investment policies on the financial markets: Evidence from the South African Boycott. *The Journal of Business, 72*(1), 35–89. https://doi.org/10.1086/209602

Tirole, J. (2017). *Economics for the common good.* https://www.torrossa.com/gs/resourceProxy?an=5560095&publisher=FZO137

Vanguard. (2025). *Vanguard's approach to responsible investment.* https://corporate.vanguard.com/content/corporatesite/us/en/corp/how-we-invest/responsible-investment.html

Wabnitz, C. C., & Blasiak, R. (2019). The rapidly changing world of ocean finance. *Marine Policy, 107*, 103526.

Wang, Y., Delgado, M. S., & Xu, J. (2023). When and where does it pay to be green? A look into socially responsible investing and the cost of equity capital. *International Journal of Corporate Social Responsibility, 8*(1), 1. https://doi.org/10.1186/s40991-023-00077-5

World Bank. (2025). *Labeled Sustainable Bonds Market Update—February 2025*. https://thedocs.worldbank.org/en/doc/cd82b4033281dab2cb1a1c71eeb691e4-0340012025/original/Labeled-Bond-Quarterly-Newsletter-Issue-No-10.pdf

Zhang, D., Zhang, Z., & Managi, S. (2019). A bibliometric analysis on green finance: Current status, development, and future directions. *Finance Research Letters, 29*, 425–430.

Ziolo, M., Filipiak, B. Z., Bąk, I., & Cheba, K. (2019). How to design more sustainable financial systems: The roles of environmental, social, and governance factors in the decision-making process. *Sustainability, 11*(20), 5604. https://doi.org/10.3390/su11205604

Zivin, J. G., & Small, A. (2005). A Modigliani-Miller theory of Altruistic corporate social responsibility. *The B.E. Journal of Economic Analysis & Policy*. https://doi.org/10.1515/1538-0653.1369

CHAPTER 3

The Rulebook: Regulatory Frameworks and Policies

Abstract This chapter examines the regulatory frameworks and policies that form the foundational "Rulebook" layer of the Sustainable Finance Operating System. It establishes regulation as essential for market stability, transparency, and directing capital toward sustainability goals. The chapter outlines the primary policy levers available: (1) altering the economics of sustainability through mechanisms like carbon pricing and incentives and (2) redirecting capital flows by mandating standardized information disclosure and classification systems. Focusing on the European Union as a key example of a comprehensive rules-based approach, the chapter details its regulatory architecture and the interconnected roles of the EU Taxonomy, the Corporate Sustainability Reporting Directive (CSRD) and the Sustainable Finance Disclosure Regulation (SFDR). It also explores the increasing role of central banks in monitoring climate-related systemic risks and integrating these into financial stability mandates. Finally, the chapter discusses the critical function of public incentives in steering investment.

Keywords EU Taxonomy · Corporate Sustainability Reporting Directive (CSRD) · Sustainable Finance Disclosure Regulation (SFDR) · Central Banks · Public incentives

W. Spiess-Knafl, *The Sustainable Finance Operating System*, Palgrave Studies in Impact Finance,
https://doi.org/10.1007/978-3-031-97499-1_3

3.1 Regulation as the Foundation

Regulations and policies are the bedrock of the Sustainable Finance Operating System. They set the core rules that ensure market stability and mandate transparency through disclosure mechanisms. This helps to reduce greenwashing and to channel capital toward genuine sustainability goals that address climate change, social inequality and responsible governance. By defining what counts as sustainable, these frameworks guarantee legal certainty for investors and issuers alike, creating level playing fields.

Before outlining specific examples, it is helpful to discuss the main types of policy levers at work globally:

- Disclosure Mandates require companies as well as financial institutions to report ESG information according to standardized templates. Examples include the EU Corporate Sustainability Reporting Directive (CSRD) for the companies and the Sustainable Finance Disclosure Regulation (SFDR) for financial intermediaries.
- Classification Systems establish taxonomies for what counts as green or sustainable activity, with the EU Taxonomy being the most ambitious to date.
- Market Mechanisms and Pricing put a direct price on externalities. Carbon pricing schemes, emissions trading systems and the EU Carbon Border Adjustment Mechanism (CBAM) incentivize emitters to reduce pollution by making it more costly to pollute.

Implementing these tools requires navigating vast and fragmented ESG data sets. This is complicated by the fact that each data set is governed by differing methodologies, disclosure formats, and quality standards. Without common definitions and reporting requirements, sustainability metrics can never be consistent and thus easily manipulated to obscure true impact. Standardized mandates, taxonomies, and pricing mechanisms bring coherence to this complexity, forcing all market participants onto the same page and thus dramatically improving data integrity, comparability and trust as a result.

Another interesting observation is that sustainable finance itself is fundamentally a social discipline. Its rules and guidelines emerge from collective human experience through expert committees, stakeholder consultations, and ongoing research. These rules and guidelines evolve as our understanding of societal risks and opportunities deepens. Rather than a static rulebook, the operating system of sustainable finance adapts over time to incorporate new insights, technologies, and social priorities. The recent Omnibus regulation which revised some of the key EU regulations around sustainability reporting is a good example. It ensured that the regulatory framework is less cumbersome, although also limiting the scope of the companies required to report (European Commission, 2025d).

Another good example of this dynamic evolution and the integration of complex social factors can be seen in the defense sector. It was long seen as incompatible with sustainable finance and classified in similar terms as gambling, tobacco and other sin stocks. Defense is associated with impact on human rights, high levels of corruption, metal pollution, and high levels of carbon emissions causing 5% of global emissions (Pedersen, 2024).

However, the Ukraine war illustrated how security can quickly become a foundational S pillar. Peace and stability are prerequisites for flourishing societies, functioning economies and the protection of human rights. That also explains how rearming Europe became a major policy objective. Defense capabilities, supplied by companies that manufacture legitimate deterrent technologies, help uphold international law and democratic values, making other ESG goals such as environmental protection or social equity achievable (Cheesley, 2025).

Yet not all defense companies are equal. ESG criteria can distinguish responsible defense firms, those that sell only to democracies bound by treaties banning cluster munitions and landmines, from those that supply offensive weapons to authoritarian regimes. Rather than blanket divestment, engaging with defense companies can also be a pathway. However, is not quite clear if all strategies are useful and compatible with ESG frameworks. One framework suggested to use non-lethal alternatives or develop products with higher precision to reduce civilian casualties (Strategy&, 2023).

3.2 The Policymaker's Toolkit: Levers for Change

Core Mechanisms Shaping the Market

Policymakers have two mutually reinforcing levers at their disposal to achieve sustainable outcomes. The first lever changes the economics of sustainability, ensuring that companies face the real costs of environmental and social impacts but also enjoy tangible benefits for sustainable practices. That means that companies have to put labels on their products or pay taxes for certain unhealthy ingredients. It can also mean that companies have access to support schemes including grants, guarantees and loans for programs which have a high value for society. The second lever redirects capital by improving the quality, comparability and availability of sustainability information and standards. This is achieved by giving capital owners and asset managers high-quality information for their investment decisions. Together, these mechanisms reshape incentives at both the project level and across entire markets and are shown in Fig. 3.1.

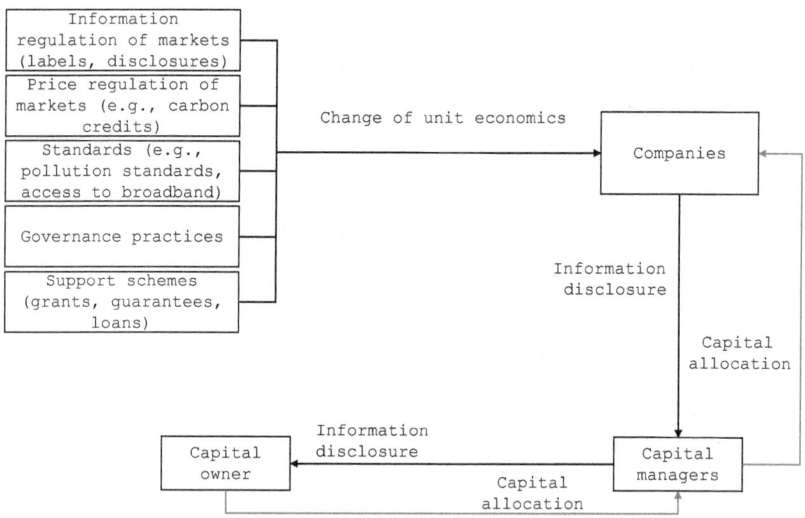

Fig. 3.1 Policy landscape. *Source* Own illustration

Lever One: Altering the Economics of Sustainability

At its heart, this lever internalizes externalities so that green activities are rewarded and brown activities incur penalties. In Europe, the Emissions Trading System sets a cap on greenhouse-gas emissions and creates a market price for allowances, while carbon taxes apply a direct levy per ton emitted. The EU's Carbon Border Adjustment Mechanism extends this logic to imports, charging for embedded carbon and prompting decarbonization abroad (European Commission, 2025a).

At the sub-national level, California's Cap-and-Trade system demonstrates how states can complement federal action with their own carbon pricing. The program started in 2013 and covers around 80 percent of California's total greenhouse gas emissions (California Air Resources Board, 2025).

Beyond prices, subsidies and positive incentives drive technology adoption. In the United States, the Inflation Reduction Act employs a range of tax credits such as Production Tax Credits and Investment Tax Credits for renewable energy, consumer rebates for home energy-efficiency upgrades (Internal Revenue Service, 2022). Large-scale funding were enabled through the Greenhouse Gas Reduction Fund and the Department of Energy's Loan Programs Office (Department of Energy, 2022; Environmental Protection Agency, 2023).

The European Green Deal also builds upon grants and tax breaks for renewable energy, efficiency measures and electric vehicles, while feed-in tariffs guarantee stable payments for wind or solar energy (European Commission, 2019). Public finance institutions further tilt the balance by supplying green loans on favorable terms and sharing risk through guarantees (e.g., European Investment Fund, 2025). Targeted support schemes such as grants for hiring people with disabilities or for sustainability-focused start-ups round out the toolkit. Collectively, these measures have marshalled trillions of euros and dollars into renewable energy, low-carbon transport and a circular economy, proving the transformative power of economic incentives. It would be difficult to imagine the current progress without these large-scale public incentives.

Lever Two: Redirecting Capital *via* Information and Standards

Economic incentives achieve their full effect only when investors trust the data and definitions underpinning sustainability claims. Mandatory disclosure rules require companies to publish consistent, detailed information on environmental, social and governance performance. In the European Union, the Corporate Sustainability Reporting Directive now covers thousands of firms, while in the United States the Securities and Exchange Commission has proposed but also faces legal challenges to climate-related financial disclosure rules. These aspects are outlined in more detail in Sect. 3.3.

At the fund level, the Sustainable Finance Disclosure Regulation compels asset managers to explain how they integrate sustainability risks and impacts into every product. Clear classification systems set a common language for what counts as sustainable and close loopholes that allow greenwashing. Market-led voluntary standards such as the Green Bond Principles guide issuers on use of proceeds, while mandatory ecolabel regulations and minimum energy performance standards for buildings ensure baseline requirements in key sectors. Regulators are also beginning to oversee the providers of ESG ratings and data, demanding transparency in their methodologies so that investors can rely on third-party assessments.

By combining rigorous disclosure, shared definitions and both voluntary and mandatory standards, this lever empowers investors to steer capital toward genuine sustainability outcomes without prescribing specific projects. The total volume which has been mobilized shows the effectiveness of this approach.

3.3 The European Union Regulatory Architecture

From Action Plan to Green Deal

In March 2018 the European Commission adopted the Sustainable Finance Action Plan in response to the High-Level Expert Group on Sustainable Finance (European Commission, 2018). This initiative laid the groundwork for new financial architecture by setting three core objectives:

Reorienting capital flows toward sustainable investment
Mainstreaming sustainability in risk management
Fostering transparency and long-termism in financial activities

Through ten concrete actions, the plan introduced key regulatory instruments that form the core of the EU's sustainable finance architecture. This architecture consists of the EU Taxonomy, foundational disclosure requirements and sustainability benchmarks which define among others what qualifies as Paris-aligned (PAB) (European Commission, 2022b).

These measures transformed how financial markets evaluate and channel capital, ensuring that sustainability considerations sit at the heart of investment decisions. It is worth repeating that there are no sanctions when investors decide to fund activities which are not considered to be sustainable.

Building on this financial foundation, the Commission launched the European Green Deal in December 2019 as its flagship strategy to achieve climate neutrality by 2050. The Green Deal extends far beyond finance, setting ambitious targets across energy systems, industry, transport, agriculture and buildings. It embeds a just transition principle, directing dedicated funding to support regions and communities most affected by the shift to a low-carbon economy. It also outlined the volume needed to be mobilized to achieve the climate goals (European Commission, 2019):

> *To achieve the ambition set by the European Green Deal, there are significant investment needs. The Commission has estimated that achieving the current 2030 climate and energy targets will require €260 billion of additional annual investment, about 1.5% of 2018 GDP. This flow of investment will need to be sustained over time. The magnitude of the investment challenge requires mobilising both the public and private sector.*

This integrated approach ensures that environmental as well as financial policy drive the same goals. By combining comprehensive regulations, incentives and disclosure standards, the EU has established itself as a global leader in sustainable finance and policy innovation. This approach also sets it apart from other jurisdictions leading to a stable and predictable regulatory environment (Bradford, 2020).

The EU Taxonomy: A Common Dictionary for Sustainability

The EU Taxonomy Regulation, in force since July 2020, establishes a common dictionary for economic activities across the European Union. As a cornerstone of the EU's sustainable finance framework, it provides clear, science-based criteria for determining whether an activity can be considered environmentally sustainable.

The Taxonomy aims to boost sustainable investment and combat greenwashing by defining which activities genuinely support environmental goals. It helps investors spot authentic green opportunities, guides companies in their transition planning and underpins credible labelling of sustainable financial products.

An activity is formally "taxonomy-aligned" only if it meets all four of the following conditions (European Commission, 2020):

(1) Substantial contribution
(2) Do no significant harm
(3) Comply with minimum safeguards
(4) Comply with the technical screening criteria

The activity must make a material positive contribution to at least one of six environmental objectives:

- Climate change mitigation
- Climate change adaptation
- Sustainable use and protection of water and marine resources
- Transition to a circular economy
- Pollution prevention and control
- Protection and restoration of biodiversity and ecosystems

While delivering on one objective, the activity must not significantly harm any of the other five. Applying DNSH criteria involves a detailed assessment of the specific Technical Screening Criteria for each of the other objectives.

The activity must respect basic social and governance standards, including international human rights and labor-rights principles (OECD guidelines, UN guiding principles, ILO conventions), as well as rules on corruption, taxation and fair competition. Delegated acts set out detailed, sector-specific performance thresholds that quantify what counts as substantial contribution and DNSH for each activity.

In general, the EU taxonomy is well accepted. About 20% of companies' capital investments are aligned with the taxonomy, while this number reaches 60% for electricity providers. In 2023, 600 European companies which already report these numbers have invested a total of €191 billion in taxonomy-aligned activities (European Commission, 2024).

Banks are also starting to provide more green loans as illustrated by the Green Asset Ratio. As argued by Frykström (2025) "it can be risky for banks to lend against collateral in the form of buildings that are at risk of flooding and other climate impacts."

The EU Taxonomy also underpins the other elements of the regulatory framework. Under the Corporate Sustainability Reporting Directive, companies must disclose the share of turnover, capital expenditure and operating expenditure linked to Taxonomy-aligned activities, using the Taxonomy's definitions to calculate these mandatory indicators.

The Sustainable Finance Disclosure Regulation uses the alignment with the Taxonomy to define "environmentally sustainable investments." Article 9 funds in this context rely on it to demonstrate sustainability as their core objective, while Article 8 funds use it to substantiate the environmental characteristics they promote.

While the EU's environmental taxonomy is established, discussions continue regarding the development of a complementary social taxonomy to define and promote investment in activities contributing to social objectives. The EU Platform on Sustainable Finance, in its February 2022 report, proposed a structure mirroring the environmental version, outlining potential objectives focused on stakeholders like "decent work," "adequate living standards and wellbeing for end-users" and "inclusive and sustainable communities."

This proposed framework would identify substantial contributions, apply "do no significant harm" criteria across social objectives and rely on international norms (Platform on Sustainable Finance, 2022). However, significant challenges remain, including defining universally applicable criteria given national differences in social legislation and concerns about administrative burden. Although the existing environmental taxonomy incorporates minimum social safeguards based on OECD and UN guidelines, the full, dedicated social taxonomy is not yet finalized and its legislative introduction appears to be currently on hold.

Corporate Sustainability Reporting Directive (CSRD)

Driven by stakeholder demands for reliable ESG information, the EU's Corporate Sustainability Reporting Directive (CSRD) represents a paradigm shift in corporate transparency. It provides a robust framework that compels larger companies to manage and disclose their environmental social and governance impacts in a consistent manner (European Commission, 2022a).

The preceding Non-Financial Reporting Directive (NFRD) introduced ESG disclosure for large public interest entities (European Commission, 2017). However, its flexible framework and limited scope led to inconsistent and incomparable data. To address these shortcomings the European Commission replaced the NFRD with the CSRD in 2022.

The new directive expands coverage to tens of thousands of companies and embeds the double materiality approach. This means evaluating both the financial risks and opportunities arising from ESG issues and the outward impacts of company operations on people and the planet. It remains to be seen how the trickle-down effect will play out as the larger companies will request their suppliers, which are often small companies to collect supply chain-related data.

At its core, the CSRD aims to furnish investors analysts and other stakeholders with standardized reliable and comparable ESG data. Mandatory European Sustainability Reporting Standards (ESRS) provide detailed guidance on disclosure metrics covering carbon emissions, water usage, labor practices, human rights and more. Companies must use XHTML and XBRL tagging so that disclosures are fully machine readable and support digital analysis. XTHML stands for (eXtensible HyperText Markup Language and is human readable, while XBRL stands for eXtensible Business Reporting Language and is machine readable (European Securities & Markets Authority, 2025). The ESRS also require firms to report alignment with the EU Taxonomy where applicable ensuring that taxonomy aligned activities are clearly identified.

CSRD's scope now covers large undertakings including non-EU companies with substantial operations in Europe. The rich corporate data generated under CSRD also forms the foundation for financial market participants' disclosures under the Sustainable Finance Disclosure Regulation helping to align product level reports with taxonomy criteria.

The specific and general objectives of the CSRD are shown in the following figure which shows the detailed level of preparation to develop this legal framework. It also hints at the importance of and the link to finance as most objectives are linked to better informed capital allocation decision-making processes.

CSRD implementation is phased by company size and scope. Firms already subject to the NFRD must publish their first CSRD compliant reports for the 2024 fiscal year with mandatory disclosures beginning in 2025. Subsequent phases extend requirements to larger corporate groups and non-EU companies operating significantly within the European Economic Area completing the rollout in later years (European Commission, 2022a) (Fig. 3.2).

Sustainable Finance Disclosure Regulation (SFDR)

The Sustainable Finance Disclosure Regulation is a cornerstone of the EU's effort to integrate sustainability into finance. It requires asset managers, institutional investors, and other market participants to explain how they incorporate environmental, social, and governance factors into investment decisions. Initially designed as a disclosure regime the SFDR, the market has been using it as a labeling framework that distinguishes three categories of funds under Articles 6, 8 and 9 (European Commission, 2023).

Enacted as Regulation EU 2019 2088 the SFDR sets uniform rules for sustainability disclosures across the financial sector. One of the key provisions include Article 4 which requires firms to disclose how they consider principal adverse impacts of investment decisions on sustainability factors such as emissions biodiversity or social issues as described in the Regulation 2019/2088 on Sustainability-related Disclosures in the Financial Services Sector:

> *Financial market participants shall publish and maintain on their websites:*
>
> (a) *where they consider principal adverse impacts of investment decisions on sustainability factors, a statement on due diligence policies with respect to those impacts, taking due account of their size, the nature and scale of their activities and the types of financial products they make available; or*
>
> (b) *where they do not consider adverse impacts of investment decisions on sustainability factors, clear reasons for why they do not do so, including, where relevant, information as to whether and when they intend to consider such adverse impacts.*

This requires financial market participants to publish these statements on their websites. Statements such as those published by Deutsche Bank

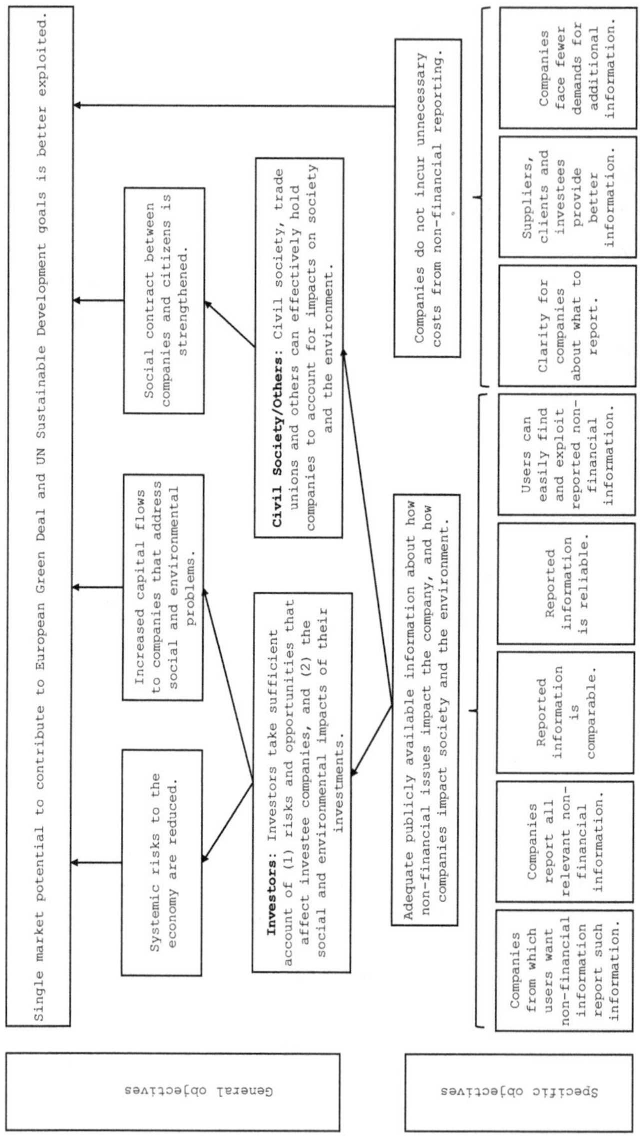

Fig. 3.2 Specific and general objectives of the Corporate Sustainability Reporting Directing (CSRD). *Source* European Commission (2021b)

(2024) contain a range of interesting data. For example, it states that 13.07% is the share of investments in investee companies with sites/operations located in or near to biodiversity sensitive areas where activities of those investee companies negatively affect those areas.

SFDR applies to a wide range of financial products from mutual funds and pension schemes to life insurance and exchange traded funds. It classifies products as follows:

- Article 6 funds provide basic disclosures on how ESG risks are integrated into investment processes
- Article 8 funds promote environmental or social characteristics. One example can be a fund marketed as ABC Green Fund that emphasizes reduced carbon emissions.
- Article 9 funds pursue sustainable investment as their main objective. One example can be XYZ Sustainable Impact Fund which allocates exclusively to high sustainability projects.

Under SFDR firms must publish comprehensive disclosures both before investment and on a regular basis detailing how they integrate ESG risks and how they assess principal adverse impacts.

However, accurate product classification is challenging due to diverse data sources and evolving interpretations of sustainability risks. Gathering high-quality data for adverse impact reporting can be complex and costly.

Since its introduction in December 2019 SFDR has been rolled out in phases. Proposed amendments aim to clarify product categorization and improve alignment with other EU regulations such as the CSRD and the EU Taxonomy (European Commission, 2025d).

An example of an Article 8 fund is the Amundi MSCI World ESG Broad Transition ETF (Amundi, 2025b). It is passively managed and aims to track the MSCI World ESG Broad CTB Select Index. This index is designed to exceed the minimum standards of the EU Climate Transition Benchmark which stipulates that the companies should have a lower carbon intensity reduction of at least 30% compared to the investable universe and a 7% year-on-year reduction of GHG intensity (Official Journal of the European Union, 2020).

The Management fees and other administrative or operating costs amount to 0.20%. As of April 2025, its top holdings included Apple (4.36%), Microsoft (4.06%), Nvidia (4.03%), Amazon.com (2.45%) and

Alphabet (1.91%), with the top ten constituents also including Meta Platforms (1.41%), Broadcom (1.20%), Tesla (1.19%), Eli Lilly & Co (1.05%) and Mastercard (0.98%). In total, these 10 companies represent approximately 23% of the fund's assets (Amundi, 2025c).

Robeco Sustainable Water serves as an example of an Article 9 fund. The fund's investment objective outlines the key themes and the screening and selection methodology (Robeco, 2025b):

> *The fund's sustainable investment objective is to help mitigate the global challenges related to scarcity, quality, and allocation of water. Water and sustainability considerations are incorporated in the investment process by the means of a target universe definition, exclusions, ESG integration, and voting. The fund only invests in companies that have a significant thematic fit as per Robeco's thematic universe methodology. Through screening on both Robeco's internally developed SDG Framework and Robeco's exclusion policy, the fund does not invest in issuers that have a negative impact on the SDGs, are in breach of international norms or where products have been deemed controversial. Financially material ESG factors are integrated in the bottom-up fundamental investment analysis to assess existing and potential ESG risks and opportunities. The integration of ESG factors in the investment analysis does not have a sustainability indicator. In addition, where a stock issuer is flagged for breaching international standards in the ongoing monitoring, the issuer will become subject to exclusion. Lastly, the fund makes use of shareholder rights and applies proxy voting in accordance with Robeco's proxy voting policy.*

It is an actively managed fund with management fees of 1.5% and service fees of 0.16%. As of April 2025, its top holdings included less well-known companies such as Agilent Technologies (4.23%), Veolia Environnement (4.22%), Xylem (4.16%), Avantor (3.89%), and Tetra Tech (3.11%). The top ten constituents also include Veralto (2.84%), Ferguson Enterprises (2.78%), Core & Main (2.72%), Waste Management (2.44%), and Halma (2.31%). In total, these ten companies represent approximately 33% of the fund's assets (Robeco, 2025a).

Connecting the Dots: The Interplay of CSRD, Taxonomy, and SFDR

The flow of sustainability information begins with the CSRD, under which companies publish standardized ESG data in line with the European Sustainability Reporting Standards (ESRS). Next, companies assess and report the extent to which their activities align with the EU Taxonomy's criteria, demonstrating substantial contribution to environmental objectives and compliance with do no significant harm requirements. Finally, asset managers draw on this reported Taxonomy alignment data to classify their products under the SFDR Articles 6, 8, and 9 framework and to provide detailed disclosures on how they integrate ESG risks and principal adverse impacts.

Corporate issuers carry the primary responsibility for accurate ESG reporting under the CSRD and for conducting the initial Taxonomy alignment assessment. Asset managers then use the reported Taxonomy alignment data from companies, applying the SFDR framework to determine which funds qualify as Article 6 (basic risk disclosure), Article 8 funds (promoting environmental or social characteristics) or Article 9 funds (focusing on sustainable investment).

This relationship between the Corporate Sustainability Reporting Directive (CSRD), EU Taxonomy and the Sustainable Finance Disclosure Regulation (SFDR) is shown in Fig. 3.3 below.

This integrated approach delivers consistent comparable data from corporate disclosures through fund level labels reducing the risk of greenwashing and directing capital toward genuine sustainability outcomes. At the same time, practitioners must address challenges such as ensuring data quality, integrating legacy IT systems and adapting to expanding Taxonomy scopes and evolving SFDR requirements.

The 2025 Omnibus Regulation: A Framework in Transition

In February 2025, the European Commission introduced the "Omnibus" legislative package, a proposal aimed at simplifying the EU's sustainable finance framework and reducing administrative burdens (European Commission, 2025f). It was presented as an effort to enhance European competitiveness given the background of high energy prices, changes in the United States, trade wars and geo-political tensions. However, the proposal has ignited a debate over whether it constitutes simplification or deregulation.

Overall, the package proposes significant amendments to the Corporate Sustainability Reporting Directive (CSRD), the EU Taxonomy Regulation and the Corporate Sustainability Due Diligence Directive (CSDDD).

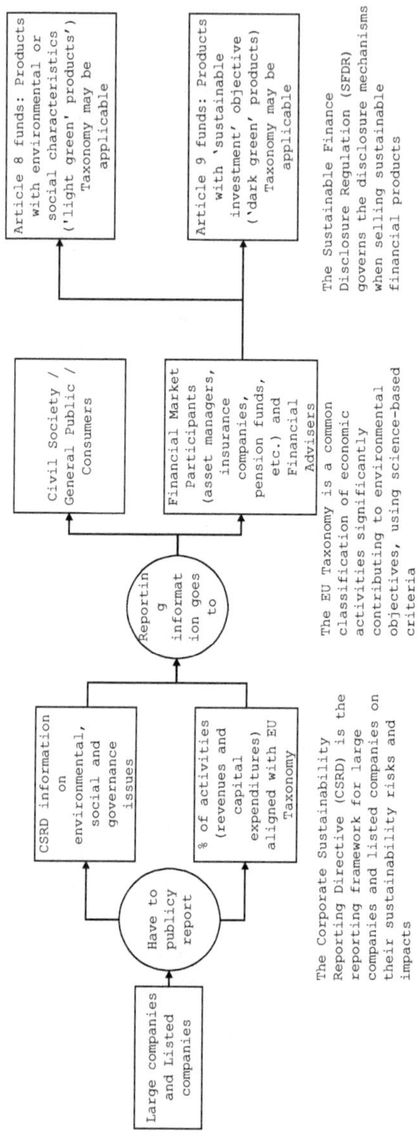

Fig. 3.3 Interplay of CSRD, EU Taxonomy and SFDR. *Source* European Commission (2021a)

The changes to the CSRD are substantial. The proposal raised the employee threshold, a move that removes approximately 80% of previously covered companies from mandatory reporting. It also postpones reporting timelines for companies not already reporting and eliminates the requirement to develop sector-specific European Sustainability Reporting Standards (ESRS). Furthermore, the option to upgrade assurance levels from "limited" to a more stringent "reasonable" standard is removed (European Parliament and the Council of the European Union, 2025a).

The CSDDD is not as relevant for the value chain of capital, but it was also covered as part of the "Omnibus package." For the CSDDD, the amendments aim to reduce the compliance burden. Due diligence obligations are narrowed, focusing primarily on a company's own operations and direct business partners, unless there is "plausible information" of risks further down the value chain. In addition, the harmonized EU-level civil liability regime is deleted, deferring to national laws and removing the threat of EU-wide penalties linked to turnover (European Parliament and the Council of the European Union, 2025b).

The EU Taxonomy is also revised to ease reporting. Taxonomy reporting will become voluntary, or "opt-in," for companies with a net turnover below a certain amount which will be fixed in the legislative process. For those who must report, the rules will be simplified. It was stated that there will be a significant reduction in required data points and an exemption for activities deemed not financially material (European Commission, 2025c).

These changes create a data gap for financial market participants who rely on CSRD data for their own disclosures under the Sustainable Finance Disclosure Regulation (SFDR). The proposals, which must still pass through the EU legislative process, signal a political recalibration of the balance between sustainability ambitions and perceived economic costs.

It is also worth noting that the instability of the policy framework has led to losses for start-ups and companies offering services around sustainability reporting. Some of them have lost a significant part of their client base and thus face heavy losses.

3.4 Central banks: The System Monitors for Stability and Risk

Historical Perspective: From Marginal Issue to Systemic Risk

Alongside regulatory bodies setting disclosure rules and standards, central banks play a crucial role as "System Monitors," monitoring the implications of climate change and sustainability for financial stability and systemic risk.

Throughout most of the twentieth century and into the early 2000s, central banks focused on price and financial stability with regional differences in priorities, with little regard for environmental or climate-related issues. Climate change was discussed in isolated policy debates and although a few economists warned of its long-term threats, these concerns remained peripheral to central banking mandates.

The financial crisis of 2008–2009 sparked a broader view of systemic risk, paving the way for central banks to consider non-traditional threats such as climate change. In 2015, Bank of England Governor Mark Carney's influential speech "*Breaking the Tragedy of the Horizon – climate change and financial stability*" brought climate risks into the mainstream (Carney, 2015).

In the speech he addressed three broad channels which affect financial stability:

(1) Physical risks impact insurance liabilities and the value of financial assets
(2) Liability risk as parties who have suffered from climate change seek compensation
(3) Transition risks on transitioning to a low-carbon economy for the value of a large range of assets.

That same year, the global momentum generated by the Paris Agreement further encouraged central banks to integrate climate considerations into their macroprudential frameworks (D'Orazio & Popoyan, 2019).

Since the late 2010s, climate considerations have become an institutional priority for central banks. In 2017, the Network for Greening the Financial System (NGFS) was established by eight central banks and supervisors and it has since expanded to include over 145 members and observers, such as the Bank of England, the European Central Bank

(ECB), the Bank of Japan and the People's Bank of China (PBoC) (Network for Greening the Financial System (NGFS), 2025). The Federal Reserve withdrew from the network citing that the network "increasingly broadened in scope, covering a wider range of issues that are outside of the Board's statutory mandate" (Federal Reserve Board, 2025).

In recent years, many central banks have implemented climate stress tests, published research on the economic impacts of climate change and adjusted their monetary policies to better account for environmental risks.

Central banks now address climate risk through several key activities: they assess climate as a systemic financial risk, conduct stress tests to gauge the resilience of banks to both physical and transition risks, adjust monetary policy tools, provide supervisory guidance and encourage transparency and disclosure. These efforts are designed to ensure that the financial system remains robust in the face of climate-related shocks while promoting a gradual shift toward sustainable investments. Fines are the last resort in this regard (Costa, 2024).

Regional Differences

European central banks have led the way in integrating climate risks into their frameworks. The Bank of England played a pivotal role in mainstreaming climate risk through its early advocacy (e.g., Carney, 2015).

The ECB has since implemented comprehensive climate stress tests. Its first economy-wide test launched in 2022 and assessed both physical and transition risks (European Central Bank, 2022). Moreover, the ECB has adjusted its asset purchase programs to favor greener issuers and is reconsidering its collateral frameworks argued the fact that around sectors with a large carbon footprint tend to have large fixed long-term capital investment needs and thus issue bonds more frequently which conflicts with market neutrality (Schnabel, 2021). As of 2025, the quantitative easing has stopped but the ECB still holds trillions of assets in the asset purchase program and the pandemic emergency purchase program (European Central Bank, 2025) In 2023, the Governing Council of the ECB decided that "for the Eurosystem's corporate bond purchases, the remaining reinvestments will be tilted more strongly towards issuers with a better climate performance" (European Central Bank, 2023). In addition, the European Banking Authority has issued guidelines that mandate the integration of ESG risks into banks' risk-management processes (European Banking Authority, 2021).

In the United States we see a different picture as the Federal Reserve followed a more cautious approach to climate measures. Political debates have contributed to this gradual engagement as the Fed joined the NGFS in 2020 and, in 2023, launched a pilot climate scenario analysis for several major U.S. banks to assess their exposure to climate impacts before leaving the NGFS again in 2025 (Federal Reserve Board, 2025) The Financial Stability Oversight Council has also emphasized climate change as an emerging threat, prompting calls for enhanced disclosure (Condon et al., 2020).

3.5 Public Incentives: Steering Capital and Driving Adoption

Public incentives are essential for redirecting capital toward sustainable investments. These incentives are a key component of the "Changing the Economics" policy lever discussed above, directly altering the financial viability and risk profile of sustainable projects.

The EU and U.S. approaches both aim to drive sustainable investments but differ in emphasis. The EU's framework is highly integrated as it combines regulatory measures and broad subsidies to mobilize capital for sustainable purposes, while the U.S. focuses on targeted tax credits, substantial loan programs and public–private partnerships with significant federal funding.

The European Union Approach: Integrated Funding Mechanisms

The EU has established a robust regulatory ecosystem to promote sustainable finance. The EU Taxonomy defines environmentally sustainable activities, while the Sustainable Finance Disclosure Regulation (SFDR) mandates transparent ESG disclosures by asset managers. These measures often lead to preferential terms and specific labels on green financial products, boosting investor confidence as outlined above.

Incentives such as the European Fund for Strategic Investments (EFSI) under the Juncker Plan (2015–2020) have mobilized around 500 billion euros in investments for renewable energy, energy efficiency, and sustainable transport and was generally found to be largely effective (European Court of Auditors, 2025). The InvestEU Program (2021–2027) and funds like the Innovation and Modernisation Funds financed by the EU Emissions Trading System offer guarantees, subsidies, and risk-sharing mechanisms that help attract private capital and de-risk new technologies (European Commission, 2025e).

The United States Approach: Tax Credits, Loans, and Partnerships

The Inflation Reduction Act of 2022 allocates hundreds of billions of dollars for clean energy and decarbonization through significant tax credits, such as Production Tax Credits and Investment Tax Credits for renewable energy and rebates for energy-efficient upgrades (Internal Revenue Service, 2022). The creation of the Greenhouse Gas Reduction Fund further supports innovative clean energy projects (Environmental Protection Agency, 2023).

The Department of Energy's Loan Programs Office, which now oversees more than $400 billion in financing, has been key in supporting early transformative projects for companies like Tesla and large-scale solar farms, boosting the development of climate-critical technologies (Department of Energy, 2022).

Municipal programs which were implemented in the 2000s were Qualified Energy Conservation Bonds and Clean Renewable Energy Bonds which provide tax-exempt, lower-cost debt financing for state and local projects in energy efficiency and renewable energy (Council of Development Finance Agencies, 2008).

The Infrastructure Investment and Jobs Act of 2021 channels significant federal funds into green initiatives like EV charging infrastructure and grid modernization (US Department of Transportation, 2021). It shows again how these initiatives have created measurable outcomes which were not imaginable without public funding.

References

Amundi. (2025a). *Amundi MSCI world ESG broad transition UCITS ETF*. https://www.amundi.co.uk/professional/product/view/IE000PB4LRO2

Amundi. (2025b). *Factsheet April 2025: Amundi MSCI World ESG broad transition UCITS ETF dist*. https://www.amundietf.com/pdfDocuments/monthly-factsheet/IE000PB4LRO2/ENG/GBR/INSTITUTIONNEL/ETF/20250430

Bradford, A. (2020). *The Brussels effect: How the European Union rules the world*. Oxford University Press.
California Air Resources Board. (2025). *Cap-and-trade program*. https://ww2.arb.ca.gov/our-work/programs/cap-and-trade-program/about
Carney, M. (2015). Breaking the tragedy of the horizon–climate change and financial stability. *Speech given at Lloyd's of London, 29*, 220–230.
Cheesley, A. (2025). *Are ESG and defence investing compatible?* https://www.wealthbriefing.com/html/article.php/are-esg-and-defence-investing-compatible--natixis-im
Condon, M., Ladin, S., Lienke, J., Panfil, M., & Song, A. (2020). Mandating disclosure of climate-related financial risk. *New York University Journal of Legislation and Public Policy, 23*, 745.
Costa, M. (2024). *ECB to fine banks that have missed climate goals*. Green Central Banking. https://greencentralbanking.com/2024/06/18/ecb-climate-fine/
Council of Development Finance Agencies. (2008). *CDFA - CDFA spotlight: Overview of Clean Renewable Energy Bonds (CREBs)*. CDFA. https://www.cdfa.net/cdfa/cdfaweb.nsf/0/07C139CF31F699468825793600063E011
Department of Energy. (2022). *Loan programs office*. https://www.energy.gov/lpo/loan-programs-office
Deutsche Bank. (2024). *Statement on principal adverse impacts of investment decisions on sustainability factors*. https://www.deutsche-bank.de/dam/deutschebank/de/shared/pdf/rechtliche-hinweise/Statement-adverse-impacts-investment-20231030.pdf
D'Orazio, P., & Popoyan, L. (2019). Fostering green investments and tackling climate-related financial risks: Which role for macroprudential policies? *Ecological Economics, 160*, 25–37. https://doi.org/10.1016/j.ecolecon.2019.01.029
Environmental Protection Agency, O. (2023, February 1). *Greenhouse gas reduction fund*. https://www.epa.gov/greenhouse-gas-reduction-fund
European Banking Authority. (2021). *EBA Report on ESG risks management and supervision*. https://www.eba.europa.eu/sites/default/files/document_library/Publications/Reports/2021/1015656/EBA%20Report%20on%20ESG%20risks%20management%20and%20supervision.pdf
European Central Bank. (2022). *2022 climate risk stress test*.
European Central Bank. (2023). *Monetary policy decisions*. https://www.ecb.europa.eu/press/pr/date/2023/html/ecb.mp230202~08a972ac76.en.html
European Central Bank. (2025, June 3). *Asset purchase programmes*. https://www.ecb.europa.eu/mopo/implement/app/html/index.en.html
European Commission. (2017). *Commission guidelines on non-financial reporting*. https://finance.ec.europa.eu/publications/commission-guidelines-non-financial-reporting_en

European Commission. (2018). *Sustainable finance: Commission's action plan for a greener and cleaner economy*. European Commission. https://ec.europa.eu/commission/presscorner/detail/en/ip_18_1404

European Commission. (2019). *Communication: The European green deal*. https://commission.europa.eu/publications/communication-european-green-deal_en

European Commission. (2020). *EU Taxonomy Navigator*. https://ec.europa.eu/sustainable-finance-taxonomy/

European Commission. (2021a). *How does the EU Taxonomy fit within the Sustainable Finance Framework?* https://finance.ec.europa.eu/system/files/2021-04/sustainable-finance-taxonomy-factsheet_en.pdf

European Commission. (2021b). *Impact assessment accompanying the document Proposal for a Directive of the European Parliament and of the Council amending Directive 2013/34/EU, Directive 2004/109/EC, Directive 2006/43/EC and Regulation (EU) No 537/2014, as regards corporate sustainability reporting (SWD(2021) 150 final)*. https://eur-lex.europa.eu/legal-content/EN/TXT/?uri=CELEX:52021SC0150

European Commission. (2022a). *Corporate sustainability reporting*. https://finance.ec.europa.eu/capital-markets-union-and-financial-markets/company-reporting-and-auditing/company-reporting/corporate-sustainability-reporting_en

European Commission. (2022b). *EU labels for benchmarks (climate, ESG) and benchmarks' ESG disclosures*.

European Commission. (2023). *Commission staff working document: Enhancing the usability of the EU Taxonomy and the overall EU sustainable finance framework*. https://eur-lex.europa.eu/legal-content/EN/TXT/?uri=CELEX:52023SC0209

European Commission. (2024). *The EU taxonomy's uptake on the ground—European Commission*. https://finance.ec.europa.eu/sustainable-finance/tools-and-standards/eu-taxonomy-sustainable-activities/eu-taxonomys-uptake-ground_en

European Commission. (2025a). *Carbon border adjustment mechanism*. https://taxation-customs.ec.europa.eu/carbon-border-adjustment-mechanism_en

European Commission. (2025b). *Commission staff working document accompanying the documents proposal for a directive of the european parliament and of the council amending directives 2006/43/EC, 2013/34/EU, (EU) 2022/2464 and (EU) 2024/1760 as regards certain corporate sustainability reporting and due diligence requirements proposal for a directive of the european parliament and of the council amending Directives (EU) 2022/2464 and (EU) 2024/1760 as regards the dates from which Member States are to apply certain *corporate sustainability reporting and due diligence requirements and specifying further steps*. https://eur-lex.europa.eu/legal-content/EN/TXT/?uri=CELEX:52025SC0080

European Commission. (2025c). *Commission staff working document related to CSRD and CSDDD*. https://commission.europa.eu/publications/omnibus-i_en

European Commission. (2025d). *InvestEU programme*. https://investeu.europa.eu/investeu-programme_en

European Commission. (2025e). *Omnibus package*. https://finance.ec.europa.eu/news/omnibus-package-2025-04-01_en

European Court of Auditors. (2025). *The European fund for strategic investments*. http://www.eca.europa.eu/en/publications/sr-2025-07

European Investment Fund. (2025). *Launch of Green Private Credit – A new solution for climate and environment financing*. https://www.eif.org/what_we_do/news/2025/france-europe-launch-of-green-private-credit-a-new-solution-for-climate-and-environment-financing.htm

European Parliament and the Council of the European Union. (2025a). *Proposal for a directive of the european parliament and of the council amending Directives 2006/43/EC, 2013/34/EU, (EU) 2022/2464 and (EU) 2024/1760 as regards certain corporate sustainability reporting and due diligence requirements*. https://eur-lex.europa.eu/legal-content/EN/TXT/?uri=CELEX:52025PC0081

European Parliament and the Council of the European Union. (2025b). *Proposal for a directive of the european parliament and of the council amending Directives (EU) 2022/2464 and (EU) 2024/1760 as regards the dates from which Member States are to apply certain corporate sustainability reporting and due diligence requirements*. https://eur-lex.europa.eu/legal-content/EN/TXT/?uri=CELEX:52025PC0080

European Securities and Markets Authority. (2025). *Electronic reporting*. https://www.esma.europa.eu/issuer-disclosure/electronic-reporting

Federal Reserve Board. (2025). *Federal reserve board announces it has withdrawn from the Network of Central Banks and Supervisors for Greening the Financial System (NGFS)*. Board of Governors of the Federal Reserve System. https://www.federalreserve.gov/newsevents/pressreleases/bcreg20250117a.htm

Frykström, N. (2025). *The green asset ratio – a metric to measure banks contribution to a green transition*. Sveriges Riksbank. https://www.riksbank.se/globalassets/media/rapporter/ekonomiska-kommentarer/engelska/2025/the-green-asset-ratio--a-metric-to-measure-banks-contribution-to-a-green-transition.pdf

Internal Revenue Service. (2022). *Inflation reduction act of 2022*. https://www.irs.gov/inflation-reduction-act-of-2022

Network for Greening the Financial System (NGFS). (2025). *Members of the network for greening the financial system*. https://www.ngfs.net/en/about-us/membership

Official Journal of the European Union. (2020, July 17). *Commission Delegated Regulation (EU) 2020/1818 of 17 July 2020 supplementing Regulation (EU) 2016/1011 of the European Parliament and of the Council as regards minimum standards for EU Climate Transition Benchmarks and EU Paris-aligned Benchmarks*. http://data.europa.eu/eli/reg_del/2020/1818/oj/eng

Pedersen, N. H. (2024). *The defence sector in focus: Common ESG risks*. PRI. https://www.unpri.org/pri-blog/the-defence-sector-in-focus-common-esg-risks/12689.article

Platform on Sustainable Finance. (2022). *Final report on social taxonomy*. https://commission.europa.eu/system/files/2022-03/280222-sustainable-finance-platform-finance-report-social-taxonomy.pdf

Regulation 2019/2088 on Sustainability-related Disclosures in the Financial Services Sector (2019). https://eur-lex.europa.eu/eli/reg/2019/2088/oj/eng

Robeco. (2025a). *Factsheet: Robeco sustainable water D EUR*. https://www.robeco.com/files/doca/CGF_WATEE_D-fact-202504-profgloben.pdf

Robeco. (2025b). *Robeco sustainable water D EUR*. https://www.robeco.com/en-int/products/funds/isin-lu2146190835/robeco-sustainable-water-d-eur

Schnabel, I. (2021). Climate change and monetary policy. *Finance & Development, 58*(3), 53–55.

Strategy&. (2023). *The evolving role of ESG in the defense industry*. https://www.strategyand.pwc.com/de/en/industries/aerospace-defense/evolving-role-esg-in-defense-industry.html

US Department of Transportation. (2021). *Infrastructure investment and jobs act*. https://www.transportation.gov/infrastructure-investment-and-jobs-act

CHAPTER 4

The Common Language: Disclosure, Materiality, and Measurement

Abstract This chapter focuses on the "Common Language" layer of the Sustainable Finance Operating System, examining the principles and practices of sustainability disclosure, materiality, and measurement. It underscores the imperative of reliable ESG disclosure as the foundation for transparency and accountability, emphasizing the need to ground reported data in the real-world context of a company. The chapter also explores the significant role of materiality in filtering relevant information, detailing the distinctions between financial, impact and double materiality perspectives and their implications under different regulatory regimes. Furthermore, the chapter outlines the key measurement frameworks and standards that provide rules for quantifying performance through specific environmental, social, and governance indicators. It addresses the persistent challenges of ESG data gaps and reliability, highlighting the importance of robust data collection and verification processes. The discussion then shifts to the practical application layer, detailing the software tools and platforms companies use for ESG data management, carbon accounting, scenario analysis, and supply chain monitoring.

Keyword Financial materiality · Double materiality · ESG data · Disclosure standards · Reporting

W. Spiess-Knafl, *The Sustainable Finance Operating System*, Palgrave Studies in Impact Finance,
https://doi.org/10.1007/978-3-031-97499-1_4

4.1 The Why of Disclosure

The Disclosure Imperative: Bedrock of the System

At the heart of sustainable finance lies a simple truth: without reliable disclosure of relevant ESG data, the entire system stops working. Just as an operating system depends on high-quality data inputs to run applications and manage resources, the sustainable finance operating system requires consistent and transparent ESG information to function effectively. Robust disclosure serves as the bedrock of this system, supplying the raw material that powers risk assessment, strategic planning and capital allocation (Achleitner et al., 2009).

The fundamental purpose of disclosure is twofold. First, to ensure transparency by making the environmental, social and governance practices of companies visible. This is the relevant perspective for investors as it helps to inform investment decisions. Second, to uphold accountability by holding those companies answerable for their impacts and performance. This is the relevant perspective for NGOs, consumers and the general public as it helps to monitor the behavior of companies. Thus, transparency shines a light on corporate actions and intentions, while accountability creates the incentive for continuous improvement and responsible behavior.

Disclosure achieves these core functions by providing investors and other stakeholders with reliable information about how companies identify, manage and report on ESG risks and opportunities. When this information flow is consistent and comparable, it reduces uncertainty, empowering decision-makers to trust in the quality of the data. In turn, that trust underpins the credibility of sustainable finance itself as it is well known in all markets where information asymmetry exists (Akerlof, 1978).

Again, the voluntary carbon market is a good illustration of how information flow helps to stabilize prices. Projects with higher quality disclosure tend to command higher prices as investors value the quality of the information.

Beyond Metrics: Grounding Disclosure in Real-World Context

While standardized disclosure is essential for comparability, effective reporting requires more than a set of numbers. Sustainability information gains real meaning only when it is grounded in a genuine understanding of the specific social and environmental challenges that surround each company.

Disclosure should reflect a company's unique interactions with its environment and its stakeholders. This means capturing both impacts and dependencies. Impacts include how operations affect local water sources, ecosystems and community health or how workforce policies influence local employment and social cohesion. Dependencies and vulnerabilities cover the reverse side of the equation: how the company relies on clean water, stable climates, or healthy communities and how it may be exposed if those systems are disrupted by resource scarcity or social unrest (e.g., Newey, 2018)

This context-specific understanding is the necessary foundation for determining materiality, which will be explored in Chapter 1.2. Without insight into the relevant social and environmental problems, a company cannot accurately judge which ESG issues pose significant financial risks or opportunities, nor identify where its own activities create meaningful positive or negative effects. In operating system terms, it is akin to identifying the critical bugs and essential processes that the data must monitor to keep the entire system running smoothly.

Grounding disclosure in this deeper understanding elevates it from a routine compliance exercise to a source of actionable insight. Stakeholders gain confidence that reported metrics reflect the issues that matter most for each company and its context, enabling smarter investment decisions, stronger stakeholder engagement and more effective strategies to address real-world challenges.

Key Drivers of Disclosure: Regulation, Markets, and Societal Expectations

Sustainability disclosure is driven by a range of forces reshaping corporate behavior. First, regulatory pressure from frameworks such as the EU's Corporate Sustainability Reporting Directive and Sustainable Finance Disclosure Regulation, alongside emerging global standards from bodies like the ISSB, require companies to report their environmental social and governance impacts in a consistent manner, enhancing comparability and reducing greenwashing (European Commission, 2022a).

Second, investors increasingly demand reliable ESG data to inform risk assessment portfolio construction and engagement. Third, societal expectations have intensified, driven by youth-led movements such as Fridays

Table 4.1 Key drivers of corporate disclosure. *Source* Own illustration

Driver	*Description*	*Example/Outcome*
Regulatory pressure	Mandated disclosure frameworks require standardized reporting	Stricter guidelines reduce greenwashing, improve comparability and lead to positive sustainability outcomes
Investor demands	Investors increasingly seek reliable ESG data to inform decisions	Firms with robust disclosures often see improved access to capital
Societal expectations	Growing public and activist pressure to report on environmental and social impacts	Activism has driven companies to adopt net-zero targets
Competitive advantage	Transparent disclosure fosters trust and reputation, translating into financial benefits	Widespread benefits such as employee attractiveness and higher customer loyalty

for Future and amplified by social media and activist campaigns, making transparent and accountable corporate conduct a social imperative and prompting many firms to adopt net zero targets and tighter supply chain oversight. It was even found that protests have an impact on share price (Ramelli et al., 2021).

Finally, companies are discovering that transparent disclosure itself can become a source of competitive advantage: by building trust with customers, attracting and retaining talent and strengthening brand reputation they often benefit from lower risk premiums and a reduced cost of capital (Table 4.1).

ESG Data Gaps and Reliability Issues

Reliable and comparable high-quality ESG data is fundamental to the sustainable finance operating system as its absence undermines transparency, accountability and credibility. Yet gathering and using such data remains a significant challenge.

One obstacle is the lack of standardization and the abundance of reporting frameworks. Varying definitions, metrics and methods make it difficult to compare performance across companies, sectors and regions.

Data gaps add to the complexity as many firms lack complete Scope 3 emissions data, struggle to quantify social impacts or operate in markets and supply chains with limited visibility. Just take the example of sourcing textiles from Bangladesh. Production can be outsourced into factories which are not monitored as intensively. In addition, cotton is hard to trace as it was observed in the case of Uzbekistan which had a history of child labor (International Labour Organization, 2022).

Even when data is available it may be inconsistent or unverified since some self-reported figures rely on estimates and independent assurance is uneven. These challenges limit informed decisions risk management and materiality assessments and can misdirect capital away from activities that are truly sustainable. However, we can expect the data quality to improve with the rollout of disclosure requirements for tens of thousands of companies.

4.2 What to Disclose and How: Materiality and Measurement Standards

Filtering the Noise: The Crucial Role of Materiality

The challenges of volume inconsistency gaps and uneven data quality highlighted in the previous section make it clear that not all sustainability information can or should be reported with equal weight. A systematic approach is needed to determine which environmental social and governance issues truly matter for each company and its stakeholders.

The materiality principle serves as that essential filter. In the context of sustainable finance materiality is used to identify and prioritize the most significant ESG issues for a specific company and its audiences. By acting as a filter, materiality separates the relevant and impactful information from the noise of countless potential data points.

This filtering is crucial because it allows companies to focus their reporting efforts on the risks and opportunities that really drive performance. Investors can then concentrate on the factors most likely to affect future value while stakeholders gain concise disclosures that highlight what really matters (Schiehll & Kolahgar, 2021).

Crucially materiality assessment must be grounded in a deep understanding of the social and environmental problems and contexts that surround the company. Without insight into potential impacts on biodiversity, human rights or community well-being and without awareness of dependences on stable climate or social stability a company cannot judge which issues are genuinely material (Garst et al., 2022).

Defining Materiality: Financial, Impact, and Double Perspectives

Materiality has traditionally referred to factors that significantly influence an investor's or creditor's economic decisions, particularly in terms of risk profiles and future cash flows. In sustainable finance financial materiality extends to ESG factors that can affect a firm's financial performance and enterprise value (e.g., Gulko et al., 2024).

The various disclosure frameworks following a single materiality approach encourage companies to assess how climate risks and opportunities might influence their financial statements and strategic plans. Critics argue this approach overlooks external impacts on the environment or society that may not immediately affect enterprise value but remain important for long-term sustainability (Oliver Yébenes, 2024).

Double materiality combines financial materiality with impact materiality. Financial materiality considers how sustainability issues influence a firm's performance and value, whereas impact materiality examines how a company's activities affect society, the environment and external stakeholders. The European Commission first introduced this concept in its climate reporting guidelines and later embedded it in the Non-Financial Reporting Directive and the Corporate Sustainability Reporting Directive. These rules require companies to disclose not only how ESG issues affect them financially but also how their operations impact people and the planet (Delgado-Ceballos et al., 2023).

Single materiality thus focuses on information that affects a company's valuation, profitability or risk profile for investors and analysts. Double materiality extends disclosure to include external impacts, serving a broader group of stakeholders such as employees, policymakers non-governmental organizations (NGOs) and the general public. Globally, regulatory approaches vary. The International Financial Reporting Standards Foundation and the International Sustainability Standards Board prioritize single materiality with an emphasis on enterprise value. In contrast, the European Union's CSRD mandate double materiality in annual reports, reflecting differing views on corporate responsibility and the role of sustainability in financial markets.

An overview of different definitions of materiality in line with the views discussed above is given in the following Table 4.2.

Table 4.2 Selected definitions of materiality. *Source* Own illustration based on Garst et al. (2022)

Organization	*Definition of materiality*
CDP	"Relevance of emissions should not be limited to sustainability topics that have a significant financial impact on your organization, or 'materiality.'" Referring to GRI definition for more guidance: "This definition puts the onus on companies to determine a materiality threshold based on internal, industry and external stakeholder interests."
CRD	"material information is any information which is reasonably capable of making a difference to the conclusions reasonable stakeholders may draw when reviewing the related information."
FASB	"Materiality is entity specific. The omission or misstatement of an item in a financial report is material if, in light of surrounding circumstances, the magnitude of the item is such that it is probable that the judgment of a reasonable person relying upon the report would have been changed or influenced by the inclusion or correction of the item."
IASB	"Information is material if omitting, misstating or obscuring it could reasonably be expected to influence decisions that the primary users of general purpose financial statements make on the basis of those financial statements, which provide financial information about a specific reporting entity."
IFRS	"Information is material if omitting it or misstating it could influence decisions that users make on the basis of financial information about a specific reporting entity."
OECD	"Material information can be defined as information whose omission or misstatement could influence the economic decisions taken by users of information."
TCFD	"Importantly, in determining whether information is material, the Task Force believes organizations should determine materiality for climate-related issues consistent with how they determine the materiality of other information included in their financial filings. In addition, the Task Force cautions organizations against prematurely concluding that climate-related risks and opportunities are not material based on perceptions of the longer-term nature of some climate-related risks."

Materiality in Practice: Regulatory Approaches and Implications

The theoretical distinction between single and double materiality has given rise to divergent regulatory and standard-setting regimes around the world. Bodies such as the IFRS Foundation and its new International Sustainability Standards Board prioritize single materiality, focusing squarely on risks and opportunities that affect enterprise value (Krivogorsky, 2024). In contrast, the European Union through the Corporate Sustainability Reporting Directive and its advisory body EFRAG (European Financial Reporting Advisory Group) mandates double materiality, requiring companies to disclose not only how sustainability issues influence their financial performance but also how their activities impact society and the environment (Dragomir et al., 2025). This divergence reflects differing global views on corporate responsibility.

For companies the applicable framework has profound practical consequences. Firms subject to double materiality rules must conduct comprehensive assessments that cover both financial and impact dimensions. They need robust data-collection systems capable of capturing metrics such as carbon emissions, water usage, labor practices, and community health indicators. Many have invested in new IT platforms and data management tools to automate reporting workflows.

External assurance engagements are usually required to validate both sets of disclosures and reinforce credibility. Internally, boards and sustainability committees must adapt governance structures to oversee these dual assessments and ensure that insights from both materiality lenses inform strategy and operations.

Moreover, asset managers consider how these approaches to materiality should be integrated in their portfolio construction and engagement strategies. Under single materiality they may consider environmental degradation only when it poses a direct threat to revenue or cost structures. By contrast, investors who consider double materiality take the company's broader social and environmental footprint into account independent of the immediate financial impact.

This wider perspective influences risk assessment models and drives more expansive stewardship activities. It consequently pushes for stronger corporate commitments on issues ranging from deforestation to human rights. As a result, the materiality lens can determine not only which companies enter its portfolio, but also how it engages with management to drive long-term sustainable value.

Measurement Frameworks and Standards: Setting the Rules

Once material topics are defined, companies rely on measurement frameworks and standards to decide what to track, how to quantify it and how to present results. These rules ensure sustainability data is consistent, comparable, reliable, and transparent across organizations and sectors.

Several bodies set the pace (Adams & Abhayawansa, 2022). The Global Reporting Initiative offers broad stakeholder-focused guidance on impact materiality. The Sustainability Accounting Standards Board provides industry-specific metrics for financially material topics and now sits within the IFRS Foundation. The International Sustainability Standards Board builds on now-disbanded TCFD as well as the SASB to create a global baseline for investor-focused disclosures. In Europe, the European Sustainability Reporting Standards, mandatory under the CSRD, require double materiality disclosure.

Quantifying Performance: Key Indicators and Impact Methodologies

Once material issues are set, companies translate them into data using common indicators and structured methodologies. Standards like ESRS, GRI, SASB and ISSB define which metrics to track and how to report them (Adams & Abhayawansa, 2022).

Environmental indicators include greenhouse gas emissions across Scopes 1, 2, and 3, water use, waste generation, and recycling rates energy efficiency as well as biodiversity measures such as deforestation. Local air quality can be measured by particulate matter levels and water quality by pollution levels (e.g., Olsthoorn et al., 2001).

Social indicators cover workforce diversity, employee turnover, health and safety incidents, training hours, community investments, and supply chain human rights issues such as child labor. Human rights in general are a good lens to assess businesses and investments (Scheck & Spiess-Knafl, 2020).

Governance indicators assess board diversity, pay linked to ESG performance, anti-corruption policies and shareholder rights among others (Kocmanová & and Šimberová, 2014). Interestingly, many of the large frauds and negative outcomes could have been avoided by using this governance indicators (e.g., Li, 2025).

Beyond raw metrics, companies use impact methodologies to map how activities lead to outcomes. Theory of Change and logic models outline steps from investments to outputs outcomes and wider effects and impacts (Spiess-Knafl & Scheck, 2023). For example, a logic model might link spending on water saving technology to lower usage and then to improved community access. Firms select approaches based on data needs and complexity to provide reliable insights that support decision-making reporting and demonstrate genuine sustainability impact.

Ensuring Data Reliability: Collection and Verification

Accurate measurement depends entirely on the reliability of the underlying data. In computer science terms, it is known as "garbage in, garbage out." Just as an operating system must validate its inputs to function correctly, the sustainable finance system relies on rigorous processes to gather and verify ESG information.

Companies use two main approaches to data collection. Primary data involves direct measurement and observation, such as monitoring facility emissions, conducting employee surveys or performing site visits. Although resource intensive, it often yields precise and context-specific results. Secondary data draws on existing sources like public databases, industry reports, third-party data providers, and supplier disclosures. This method can be more efficient but may lack the specificity of primary collection. For example, the carbon footprint of a factory might be different than the industry benchmark.

Data collection must be followed by verification to guarantee integrity. Internal controls is a first step applying standardized procedures data validation rules and regular reconciliations. Independent verification or assurance by external organizations adds credibility and helps meet regulatory requirements such as those under the CSRD. Firms may seek limited or reasonable levels of assurance depending on stakeholder needs.

Technological aids for verification include data management platforms with built-in validation checks and emerging applications of distributed ledger technology to secure supply chain records. The goal of these protocols is to minimize errors, ensure consistency, and build stakeholder trust in the accuracy of ESG reporting.

4.3 Tools for Companies: The Applications Layer of the Sustainable Finance OS

Software as the Interface for Sustainability Data

Effective sustainability management means that companies have to use advanced software tools. This layer functions as the interface that transforms raw ESG data into actionable insights, ensuring regulatory compliance and driving strategic decision-making to support sustainable growth. As companies face increasing regulatory requirements such as the EU's CSRD and heightened investor demands for transparency, these tools serve as the bridge between comprehensive data collection and insightful reporting.

The regulation is applicable for larger and listed companies, but it can be expected that small and medium enterprises which are suppliers to large companies have to follow the same procedures. It is often referred to as trickle-down effect and can be expected to happen in the B2B area. Examples might be timber companies, electronic suppliers, IT providers or catering companies which deliver their products to larger and listed companies.

ESG Data Management & Reporting Platforms

Centralized reporting platforms are the backbone of the sustainability measurement process, serving as the control center where vast ESG data streams are integrated and managed. These platforms consolidate information from disparate sources, streamline workflows and generate reports that comply with global standards. Workiva, for example, offers a cloud-based solution that integrates financial and non-financial data, supporting regulatory filings with features like XBRL tagging. It has more than 6,300 clients using its services.

SAP Sustainability Control Tower takes advantage of ERP integration with its other products merging financial, operational and sustainability data. Wolters Kluwer Enablon offers comprehensive modules focused on environmental, health and safety management alongside broader ESG functions. It shows that these tools enable companies to handle multiple frameworks concurrently and achieve audit-ready outputs. However, all of them have different strengths which need to be considered when selecting one tool.

Carbon Accounting and GHG Footprinting Tools

Accurate carbon accounting is a critical component of sustainability reporting, acting as a specialized calculation engine within the operating system. Tools like Sphera facilitate detailed lifecycle assessments and environmental data tracking, ensuring that corporate emissions align with the GHG Protocol and other international standards. Sphera manages a dataset of 20,000 life cycle assessments (LCAs).

Normative defines itself as "carbon accounting engine" and automates the calculation of Scope 1, 2, and 3 emissions by leveraging financial transaction data and standardized emission factors, thereby streamlining the reporting process for organizations of all sizes. Persefoni focuses on providing portfolio-level carbon analytics for both corporate and financial institution use, with advanced scenario modeling capabilities that align with TCFD recommendations.

In general, these tools offer robust insights into carbon performance, enabling firms to set targets and minimize their environmental footprint. While these tools offer sophisticated analysis, it might be enough in many cases to follow simple procedures. For example, some smaller banks only use general sector averages when calculating the carbon footprint of their loan portfolio. This is based on the understanding that averages are sufficient given the large numbers of clients and their relatively small carbon footprint.

Scenario Analysis and ESG Risk Modeling Tools

The preparation for uncertain future outcomes is done with scenario analysis tools which predict the impacts of climate change and other ESG risks. These tools act as the system's forward-looking risk modules, allowing companies to simulate stress tests and quantify potential impacts under various scenarios.

The MSCI ESG Manager has a Climate Value-at-Risk module which integrates ESG data with climate risk metrics to evaluate the financial implications of different temperature pathways. Meanwhile, the open-source tool PACTA (Paris Agreement Capital Transition Assessment) helps financial institutions assess the alignment of their portfolios with the goals of the Paris Agreement. Again, open-source approaches have a higher value to the community as it can be better assessed, but there are limits given the need to process large volume of data and develop appropriate models.

Supply Chain and Stakeholder Engagement Tools

Sustainability extends beyond a company's direct operations to its entire value chain and sophisticated tools are needed to manage this complex network. Large companies have thousands of suppliers across different value chains in dozens of jurisdictions. Disruptions during the Pandemic and trade conflicts have shown the vulnerability of supply chains (e.g., Moosavi et al., 2022).

Supply chain and stakeholder engagement platforms facilitate the collection and analysis of ESG data across diverse suppliers and external partners. EcoVadis provides detailed sustainability ratings for suppliers by evaluating aspects such as environmental impact, labor practices, and ethics, thereby enabling companies to mitigate risks across their supply chains. The company states that they have screened more than 3 million companies and rated more than 150,000 companies. Datamaran employs artificial intelligence to continuously scan external information sources such as media, regulatory updates, and peer disclosures among other product offers. This helps companies to identify and prioritize material ESG issues.

In addition, emerging blockchain solutions are beginning to enhance traceability and ensure immutable verification of sustainability claims, which further strengthens data integrity and transparency throughout the supply chain (Spiess-Knafl, 2022).

Implementation Challenges

The preceding sections detailed the various software tools that form the applications layer of the Sustainable Finance Operating System ranging from comprehensive ESG data management platforms to specialized carbon accounting and risk modeling applications.

However, their effective deployment within organizations presents considerable practical difficulties outlined below:

- Data integrity and methodological challenges: Organizations face significant hurdles in sourcing reliable and comparable ESG data, particularly for non-financial metrics like Scope 3 emissions often reliant on incomplete internal records. In addition, there is always the inherent tension between the need of tools to follow standardization paths with the specific needs of companies.

- Data security and privacy concerns: Some ESG data concerns sensitive information around human rights, employee data, safety procedures and personal information. Detailed information might also be a risk for cyber security.
- Technological integration and security concerns: Technical integration barriers arise from linking specialized sustainability platforms with legacy enterprise systems. These systems are often based on large and heterogeneous data sets.
- Organizational adoption and strategic alignment: Successful implementation is frequently impeded by organizational factors, requiring strong leadership commitment and governance. In addition, there is a need for cross-functional collaboration among finance, sustainability, and operations developments to overcome internal resistance to.
- Navigating Regulatory Complexity: Navigating the constantly evolving and often fragmented landscape of regulatory mandates demands significant organizational agility.

4.4 Sustainability in Practice: Comparative Corporate Case Studies

What Companies Typically Disclose

Sustainability reporting is the practice of disclosing both financial and ESG data to give stakeholders a comprehensive view of a company's performance and risk management. Larger firms publish them annually, although the scope and depth vary by industry jurisdiction and the company's materiality choices.

Environmental disclosures focus on a company's interactions with the natural environment. Common metrics include greenhouse gas emissions by Scope 1, Scope 2, and Scope 3 sources, energy consumption and efficiency, water withdrawal and use, waste generation and recycling rates and biodiversity impacts such as land use change or deforestation. Local indicators like air quality or water quality often supplement these core figures.

Social disclosures cover impacts on people and communities. Typical topics include labor practices such as wages working conditions and freedom of association workforce diversity and inclusion statistics, employee health and safety incident rates and training hours, community engagement and investment, and human rights performance in the supply chain, including measures to prevent child labor. In some sectors, companies also report school attendance or caregiver employment metrics to illustrate social outcomes in their areas of operation.

Governance disclosures address systems of oversight control and accountability. Common elements are board composition including diversity and independence, executive compensation structures with links to ESG performance, risk management policies, business ethics and anti-corruption measures and shareholder rights and engagement processes.

Reporting also goes beyond quantitative metrics. Narrative disclosures explain the company's sustainability strategy goals, challenges, risk management and future objectives providing crucial context for the data. Integrated Reporting combines financial and non-financial information in a single report to demonstrate their interconnectedness and the impact of ESG performance on overall corporate value creation.

While these components form the backbone of most sustainability reports how companies select measure, prioritize and present information varies widely based on their industry strategy and materiality assessment.

One of the best-known examples for sustainable corporate activities is Patagonia. Patagonia's approach to Corporate Social Responsibility (CSR) and Environmental, Social, and Governance (ESG) is deeply ingrained in its core philosophy and operations (Chouinard, 2016).

On the ownership structure, 100% of the company's voting stock is held by the Patagonia Purpose Trust and 100% of the non-voting stock is held by the Holdfast Collective which also receives the dividends to defend nature. The founder has called it the "Earth as only shareholder" ownership structure. In addition, 1% of annual sales are given away for purposes close to its mission (Patagonia, 2022).

Key activities include a strong focus on sustainable materials, aiming to eliminate virgin petroleum and increase recycled content. One example is the program called "Worn Wear" for product longevity. The company champions fair labor through its extensive fair trade program, benefiting over 85,000 workers, and is actively working toward living wages in its apparel factories. Environmental activism is central, with 1% of sales donated to grassroots environmental groups via 1% for the Planet and community engagement fostered through Patagonia Action Works (Patagonia, 2025).

Patagonia evaluates and monitors its ESG performance through a combination of third-party certifications, internal tracking, and external scrutiny. One example is the B Corp certification which serves as a holistic benchmark, with Patagonia achieving a high score of 166.0 in its 2023 recertification, reflecting strong performance, particularly in community and environmental impact (B Lab, 20,224).

Comparing Sustainability Performance across Companies

Understanding how companies measure as well as report sustainability is essential for the analysis of the sustainable finance operating system. In this context, sustainability reporting is the critical filtering mechanism that converts raw ESG data into actionable insights. A larger amount of ESG disclosure can even lead to greater ESG rating disagreement (Christensen et al., 2022).

The examples below are pairs from the same industry to understand the differences in how they approach sustainability. In general, the differences across the ESG scores show the difficulties in comparing companies (Berg et al., 2022) (Table 4.3).

Each industry has its own challenges. The products of the automotive industry are responsible for a large share of global CO2 emissions. For example, road transport is responsible for a fifth of the total EU emissions (European Parliament, 2019). The main pathway to reduce the carbon footprint is to increase the share of electric vehicles. Smaller efficiency

Table 4.3 ESG Scores for companies in different industries. *Source* Own illustration based on KnowESG, 2025; S&P Global, 2025; Sustainalytics, 2025

Company	*MSCI ESG rating (AAA–CCC)*	*Sustainalytics ESG risk (score, category)*	*S&P Global ESG score (100-pt scale)*
General motors	BBB	27.6 (Medium Risk)	49
Ford	BB	27.6 (Medium Risk)	37
Microsoft	AAA	17.4 (Low Risk)	53
Alphabet	BBB	24.9 (Medium Risk)	45
Ryanair	A	21.0 (Medium Risk)	45
Lufthansa	AA	26.0 (Medium Risk)	48

gains can be captured by improving production processes or improving the fuel efficiency of combustion engines. Interestingly, General Motors and Ford have the same Sustainalytics ESG Risk Score of 27.6. Reports and news articles mainly discuss the transition to a state where every car is an electric vehicle. Other companies such as Tesla do not necessarily outperform these companies as they have a better profile in terms of the carbon footprint but might have issues with governance or social topics.

Large tech companies are commonly known to have a large carbon footprint which will likely increase given the large computing requirements of generative AI applications (Stocker et al., 2024). Two of the leading companies are Microsoft and Alphabet. Microsoft outperforms Alphabet in all relevant ESG ratings and has the highest rating in the MSCI ESG. The reports usually focus on carbon reduction commitment and the approaches to social issues. Alphabet has strong policies on climate change and resource efficiency but is more vulnerable for ESG controversies around user privacy and data protection as well as content governance challenges for YouTube.

The aviation industry is responsible for 2.5% of global energy-related CO_2 emissions and growing faster than rail, road or shipping according to numbers from (International Energy Agency, 2025). Industry reports stress aspects such as new fuel, carbon offsets, better scheduling, or reduced plastic waste on board. Climate-wise, reducing absolute emissions is difficult to achieve.

The ESG statements by Ryanair are interesting as it shows the realities in this field (Ryanair, 2025).

> *During Q3, MSCI reconfirmed Ryanair's 'A' rating, we retained Sustainalytics No.1 global large cap airline ESG ranking and Ryanair became the first major airline to have its environmental targets (to reduce CO2 per pax/km by 29% to c.50grams by 2031) validated to the latest SBTi guidelines. In Q3 the retro-fit of winglets to our B737NG fleet (target of 409 by 2026) continued, reducing fuel burn by 1.5% and noise by 6%, and we took delivery of 2 Gamechangers (4% more seats, 16% less fuel & CO2). Our new aircraft, increasing use of winglets and SAF commitments positions Ryanair as one of the EU's most environmentally efficient airlines. Plans to migrate the remaining 25% of customers who don't already check-in via the Ryanair App to paperless boarding during 2025 are progressing well. This initiative will remove pprox.. 300 tonnes of paper annually and will ensure that all customers have access to Day of Travel updates, live flight information,*

the convenience of Order to Seat for onboard purchases and the many other features contained in the Ryanair App (the ideal travel companion)

All initiatives like high passenger numbers, lower fuel consumption or the focus on the app use of passengers are reasonable business objectives. It is not entirely clear how a similar airline might achieve a different ESG score given very similar challenges. The small differences in the ESG scores show the similarities of the business models.

References

Achleitner, A.-K., Bassen, A., Roder, B., & Spiess-Knafl, W. (2009). *Reporting in social entrepreneurship.*

Adams, C. A., & Abhayawansa, S. (2022). Connecting the COVID-19 pandemic, environmental, social and governance (ESG) investing and calls for 'harmonisation' of sustainability reporting. *Critical Perspectives on Accounting, 82*, Article 102309. https://doi.org/10.1016/j.cpa.2021.102309

Akerlof, G. A. (1978). The market for "lemons": Quality uncertainty and the market mechanism. In *Uncertainty in economics* (pp. 235–251). Elsevier. https://www.sciencedirect.com/science/article/pii/B978012214850750022X

B Lab. (20224). *Patagonia—certified B corporation—B Lab Global.* https://www.bcorporation.net//en-us/find-a-b-corp/company/patagonia-inc/

Berg, F., Kölbel, J. F., & Rigobon, R. (2022). Aggregate confusion: The divergence of ESG ratings. *Review of Finance, 26*(6), 1315–1344.

Chouinard, Y. (2016). *Let my people go surfing: The education of a reluctant businessman–Including 10 more years of business unusual.* Penguin.

Christensen, D. M., Serafeim, G., & Sikochi, A. (2022). Why is corporate virtue in the eye of the beholder? The case of ESG ratings. *The Accounting Review, 97*(1), 147–175. https://doi.org/10.2308/TAR-2019-0506

Delgado-Ceballos, J., Ortiz-De-Mandojana, N., Antolín-López, R., & Montiel, I. (2023). Connecting the sustainable development goals to firm-level sustainability and ESG factors: The need for double materiality. *BRQ Business Research Quarterly, 26*(1), 2–10. https://doi.org/10.1177/23409444221140919

Dragomir, V. D., Dumitru, M., Chersan, I. C., Gorgan, C., & Păunescu, M. (2025). Double materiality disclosure as an emerging practice: The assessment process, impacts, risks, and opportunities. *Accounting in Europe, 22*(1), 103–140. https://doi.org/10.1080/17449480.2024.2339264

European Commission. (2022). *Corporate sustainability reporting*. https://finance.ec.europa.eu/capital-markets-union-and-financial-markets/company-reporting-and-auditing/company-reporting/corporate-sustainability-reporting_en

European Commission. (2022a). *Corporate sustainability reporting*. https://finance.ec.europa.eu/capital-markets-union-and-financial-markets/company-reporting-and-auditing/company-reporting/corporate-sustainability-reporting_en

European Parliament. (2019). *CO_2 emissions from cars: Facts and figures*. https://www.europarl.europa.eu/topics/en/article/20190313STO31218/co2-emissions-from-cars-facts-and-figures-infographics

Garst, J., Maas, K., & Suijs, J. (2022). Materiality assessment is an art, not a science: Selecting ESG topics for sustainability reports. *California Management Review, 65*(1), 64–90. https://doi.org/10.1177/00081256221120692

Gulko, N., Gerardou, F. S., & Withanage, N. (2024). The concept of materiality in CSR and SDGs reporting: definitions, interpretation, application, and sustainable value creation. In *The emerald handbook of ethical finance and corporate social responsibility: A framework for sustainable development* (pp. 1–30). Emerald Publishing Limited. https://www.emerald.com/insight/content/doi/https://doi.org/10.1108/978-1-80455-406-720241001/full/html

International Energy Agency. (2025). *Aviation*. IEA. https://www.iea.org/energy-system/transport/aviation

International Labour Organization. (2022, March 1). *Uzbek cotton is free from systemic child labour and forced labour*. https://www.ilo.org/resource/news/uzbek-cotton-free-systemic-child-labour-and-forced-labour

KnowESG. (2025). *MSCI, LSEG, sustanalytics ESG ratings*. KnowESG. https://www.knowesg.com

Kocmanová, A., & Šimberová, I. (2014). Determination of environmental, social and corporate governance indicators: Framework in the measurement of sustainable performance. *Journal of Business Economics and Management, 15*(5), 1017–1033. https://doi.org/10.3846/16111699.2013.791637

Krivogorsky, V. (2024). Sustainability reporting with two different voices: The European Union and the International Sustainability Standards Board. *Journal of International Accounting, Auditing and Taxation, 56*, Article 100635. https://doi.org/10.1016/j.intaccaudtax.2024.100635

Li, J. (2025). Corporate governance, fraud learning cycles, and financial fraud detection: Evidence from Chinese listed firms. *Research in International Business and Finance, 76*, Article 102832. https://doi.org/10.1016/j.ribaf.2025.102832

Moosavi, J., Fathollahi-Fard, A. M., & Dulebenets, M. A. (2022). Supply chain disruption during the COVID-19 pandemic: Recognizing potential disruption management strategies. *International Journal of Disaster Risk Reduction, 75*, Article 102983. https://doi.org/10.1016/j.ijdrr.2022.102983

Newey, L. R. (2018). 'Changing the system': Compensatory versus transformative social entrepreneurship. *Journal of Social Entrepreneurship, 9*(1), 13–30. https://doi.org/10.1080/19420676.2017.1408671

Oliver Yébenes, M. (2024). Climate change, ESG criteria and recent regulation: Challenges and opportunities. *Eurasian Economic Review, 14*(1), 87–120. https://doi.org/10.1007/s40822-023-00251-x

Olsthoorn, X., Tyteca, D., Wehrmeyer, W., & Wagner, M. (2001). Environmental indicators for business: A review of the literature and standardisation methods. *Journal of Cleaner Production, 9*(5), 453–463. https://doi.org/10.1016/S0959-6526(01)00005-1

Patagonia. (2022). *Yvon chouinard donates patagonia to fight climate crisis*. https://eu.patagonia.com/gb/en/ownership/

Patagonia. (2025). *Social responsibility*. https://eu.patagonia.com/gb/en/social-responsibility/

Ramelli, S., Ossola, E., & Rancan, M. (2021). Stock price effects of climate activism: Evidence from the first Global Climate Strike. *Journal of Corporate Finance, 69*, Article 102018. https://doi.org/10.1016/j.jcorpfin.2021.102018

Ryanair. (2025). *Ryanair Q3 2025 statements*. https://investor.ryanair.com/wp-content/uploads/2025/01/Q3-FY25-Ryanair-Results.pdf

Scheck, B., & Spiess-Knafl, W. (2020). *Impact investing in the framework of business and human rights*. European Parliament. https://www.europarl.europa.eu/thinktank/de/document.html?reference=EXPO_IDA(2020)603490

Schiehll, E., & Kolahgar, S. (2021). Financial materiality in the informativeness of sustainability reporting. *Business Strategy and the Environment, 30*(2), 840–855. https://doi.org/10.1002/bse.2657

S&P Global. (2025). *ESG scores: Rating company ESG performance*. https://www.spglobal.com/esg/solutions/esg-scores-data

Spiess-Knafl, W. (2022). *Artificial intelligence and blockchain for social impact: Social business models and impact finance*. Taylor & Francis Ltd.

Spiess-Knafl, W., & Scheck, B. (2023). *Impact investing: Instruments, mechanisms and actors* (2nd ed.). Palgrave Macmillan.

Stocker, V., Mariotte, N., Ullrich, A., & Rejeski, D. (2024). *ICT sustainability reporting strategies of large tech companies: Changes in format, scope, and content* (SSRN Scholarly Paper No. 4927128). Social Science Research Network. https://papers.ssrn.com/abstract=4927128

Sustainalytics. (2025). *Company ESG risk rating*. Sustainalytics.Com. https://www.sustainalytics.com/esg-rating

CHAPTER 5

Financial Instruments

Abstract This chapter explores the diverse range of financial instruments that function as the operational "Applications Layer" within the Sustainable Finance Operating System, translating rules, standards and data into directed capital flows. It classifies these instruments first by their intended purpose and examines the pathways capital follows. The chapter then details foundational sustainable investment techniques such as Socially Responsible Investing (SRI), ESG Integration and Impact Investing, highlighting their distinct motivations and approaches. Furthermore, the chapter analyzes instruments based on their financial structure, discussing equity strategies in public and private markets, various debt instruments like green, social, sustainability and sustainability-linked bonds, and loans as well as innovative structures like blended finance and market-based mechanisms such as carbon markets. It addresses the ongoing debate regarding the financial returns of sustainable finance, exploring the link between ESG factors and corporate or portfolio performance as well as the fundamental considerations in valuing investments that incorporate non-financial objectives alongside financial ones.

Keywords Financial instruments · Socially responsible investing · ESG integration · Impact investing · Financial returns

W. Spiess-Knafl, *The Sustainable Finance Operating System*, Palgrave Studies in Impact Finance,
https://doi.org/10.1007/978-3-031-97499-1_5

5.1 Financial Instruments for Sustainable Finance: The Applications Layer

Thinking of financial instruments as applications on the Sustainable Finance Operating System helps clarify their role. They are the specific tools that use the system's underlying layer of data, regulations, and standards to perform distinct tasks, namely directing capital toward sustainable outcomes. To understand this diverse range of applications, this chapter will explore them through two primary lenses: first by their intended purpose and impact pathway and then by their underlying financial structure.

The core purpose of these instruments is to combine financial returns with environmental and social goals with varying degrees in how each dimension is prioritized. Green bonds fund renewable energy projects, sustainability-linked loans tie pricing terms to measurable ESG performance and blended finance vehicles leverage public or philanthropic capital to de-risk investments and attract private funds. Other examples include social bonds that support affordable housing and climate impact funds that channel investment into nature-based solutions.

A significant tension shapes the design of these instruments. There is a difference between pursuing financial value by optimizing risk-adjusted returns using ESG data as an input and upholding broader societal values, where positive impact and ethical considerations are the primary objective. This is the frame of "value" vs. "values" discussed above.

Some applications are built to deliver market-competitive returns with ESG considerations framed as risk and opportunity factors, while others are coded explicitly for impact, accepting lower financial returns in exchange for greater social or environmental benefit. Understanding this duality is key to selecting the right instrument and assessing its true contribution, guarding against potential bugs such as greenwashing.

As the Sustainable Finance Operating System continues to evolve, so too will the innovation in structuring these financial applications. For example, social impact bonds have only been launched in 2010 (Social Finance, 2017). The use of data to assess social and environmental outcomes has only started to be done more widely in 2010 and the use of artificial intelligence in the early 2020s.

5.2 Classifying by Intent: Purpose and Impact Pathways

Categories and Capital Flow Pathways

Sustainable finance instruments are commonly categorized based on their primary sustainability objectives and the types of impact they aim to achieve. These key categories include:

- Climate mitigation instruments focus on reducing greenhouse gas emissions. Examples include green bonds for renewable energy projects, carbon credits and climate-focused venture capital that funds low-emission technology.
- Climate adaptation instruments aim to enhance resilience to climate impacts. Examples include catastrophe bonds for disaster risk financing, resilience bonds that fund flood protection and blended finance structures supporting climate-proof infrastructure.
- Social impact instruments address societal challenges by improving access to essential services or reducing inequality, often prioritizing values-based outcomes alongside financial considerations. Examples include social bonds for affordable housing, impact funds for health and education and microfinance for financial inclusion.
- Biodiversity and natural capital instruments target the goals of preserving and restoring ecosystems. Examples include blue bonds for ocean conservation, sustainable forestry funds and biodiversity offsets that support habitat protection and restoration.
- Multi-objective instruments integrate several sustainability goals simultaneously. Examples include sustainability bonds and SDG-linked instruments that channel capital toward projects with both environmental and social benefits.

Understanding Capital Flow Pathways

Beyond their purpose, it is useful to understand the different pathways through which these instruments allocate capital. These pathways operate across the categories defined above. Three primary pathways can be identified:

- Direct Funding: This pathway involves capital flowing directly to designated sustainable projects or activities. Use-of-proceeds instruments exemplify this approach. This happens where funds raised through green or social bonds are explicitly earmarked for eligible environmental or social initiatives. This ensures a clear link between the financing and specific sustainable undertakings.
- Indirect Funding: This pathway operates through market signals and incentives that influence an issuer's overall cost of capital or financial terms based on sustainability characteristics or performance. Mechanisms include increased investor demand for entities with strong ESG profiles potentially leading to lower funding costs (e.g., via oversubscription or a "greenium" on bonds), serving as a market signal. Performance-linked funding operates similarly, where loan interest rates or bond coupons are adjusted based on the issuer achieving predefined ESG targets, indirectly incentivizing broader sustainability improvements.
- Risk Transfer: This pathway utilizes financial instruments designed to reallocate specific sustainability-related risks, particularly those associated with climate change or environmental events, from entities less able to bear them to those better equipped, such as insurers or specialized investors. Catastrophe bonds, which cover losses from natural disasters, or climate derivatives are examples of this mechanism.

Below is a summary table that categorizes sustainable finance instruments by purpose and impact pathway, along with examples and descriptions (Table 5.1).

Key Stakeholders and Their Roles

The sustainable finance ecosystem encompasses a diverse network of participants, each fulfilling a distinct role within the operating system to channel capital and information toward ESG objectives.

Issuers and recipients occupy one of the central nodes. Corporations, for instance, issue green bonds and secure sustainability-linked loans to finance projects and signal their ESG commitments. Large companies can access a wide array of instruments directly, while small and medium-sized enterprises rely on specialized intermediaries to reach capital markets. Banks underwrite and issue green bonds, structure

Table 5.1 Different instruments

Category	*Primary objective & impact pathway*	*Examples*	*Market Maturity Overview (2024)*
Climate mitigation instruments	Reduce greenhouse gas emissions through technology deployment and infrastructure development	Green bonds for renewable energy, carbon credits, climate VC	Annual issuance ~ US$1 trillion
Climate adaptation instruments	Enhance resilience by financing projects that reduce vulnerability to climate risks	Catastrophe bonds, resilience bonds, adaptation-focused blended finance	Emerging segment with growing interest, particularly in areas hit by extreme weather
Social impact instruments	Improve access to essential services and reduce inequality	Social bonds, impact funds, microfinance solutions	Part of the US$1.57 trillion impact investing market
Biodiversity & natural capital instruments	Preserve and restore ecosystems through sustainable resource management	Blue bonds, sustainable forestry funds, biodiversity offsets	Early stage but rapidly expanding as demand for nature-based solutions increases
Multi-objective instruments	Address multiple sustainability dimensions simultaneously	Sustainability bonds, SDG-linked instruments	Evolving segment integrating diverse ESG factors across various sectors

Source Own illustration

blended finance arrangements and extend sustainability-linked facilities, acting both as financiers and distribution channels. Public sector entities from supranational entities such as the European Union to regional municipalities in California issue sovereign and municipal green bonds to fund infrastructure and environmental programs.

Development banks further structure concessional financing and guarantees to mobilize private capital. Project developers in renewable energy, sustainable transport, and other green sectors depend on project finance, green bonds and dedicated impact funds to secure necessary funding.

Complementing the issuers are the investors and capital providers who allocate funds. Institutional investors such as pension funds, insurance companies and sovereign wealth funds deploy long-term capital with ESG integration embedded in their mandates. Asset managers design and oversee ESG-integrated funds, thematic vehicles and impact portfolios, bridging the gap between asset owners and capital recipients. Banks bolster this system by underwriting public offerings and arranging sustainability-linked facilities. Individual investors participate through retail sustainable funds, digital investment platforms and even crowdfunding schemes targeting local sustainability initiatives.

The effective functioning of issuers and investors relies upon the existence of an enabling layer, comprising data providers, regulators, verifiers and advisors. ESG rating agencies and data vendors analyze corporate. Independent assurance providers conduct third-party verification of sustainability claims, which is important for lending credibility to labeled instruments like green bonds and sustainability-linked loans.

5.3 Foundational Techniques: Core Sustainable Investment Approaches

Socially Responsible Investing (SRI)

Socially Responsible Investing is the foundational approach within sustainable finance. It aligns portfolios with an investor's ethical moral or religious values rather than seeking specific environmental or social outcomes. The core motivation is non-financial and relies primarily on negative screening to exclude sectors or companies deemed harmful or unethical. Typical sectors are tobacco, weapons, fossil fuel extraction, or gambling. In some cases, investors apply positive screening by selecting companies with strong practices or adhere to norms such as Sharia-compliant criteria. The primary goal is to ensure the portfolio reflects the investor's values rather than to optimize financial returns through sustainability factors (Berry & Junkus, 2013).

SRI strategies which are based on exclusionary screening remain a substantial segment of public capital markets. Exact figures vary as methodologies overlap with broader ESG approaches, but the Global Sustainable Investment Alliance review published in 2023 shows that exclusionary screening was responsible for 77% of all volume in Europe and features prominently worldwide. Although total sustainable investment assets dipped slightly between 2020 and 2022 due to market fluctuations and methodological changes, investor demand for values-aligned portfolios endures.

A typical example might be an SRI investor who excludes all fossil fuel companies from a fund regardless of those companies' transition plans or financial health. A faith-based pension scheme may similarly bar investments in alcohol or gaming businesses to honor its guiding principles.

ESG Integration

Another approach is ESG Integration, which actively embeds financially material environmental, social and governance factors into traditional financial analysis and investment decision-making. It recognizes that issues such as carbon emissions, water management, labor practices and board composition can directly influence enterprise value and long-term cash flows. Therefore, this approach treats ESG data as a vital input for risk assessment and valuation models (Eccles & Klimenko, 2019).

Key ESG integration strategies include:

- Systematic analysis: Incorporating ESG metrics such as greenhouse gas emissions, water usage, labor standards, board diversity and executive pay into company valuations, credit models and portfolio risk reviews.
- Positive screening or best in class: Selecting firms that outperform peers on material ESG indicators within their industry.
- Thematic investing: Targeting opportunities tied to ESG themes like clean energy, water innovation, sustainable agriculture or healthcare advancement.
- Active ownership and engagement: Using shareholder influence to press companies for better ESG practices and disclosures.

This approach serves two main purposes: it mitigates risks such as regulatory penalties, reputational harm and stranded assets and it uncovers opportunities such as operational efficiencies innovation and stronger stakeholder relationships. Ultimately, ESG Integration seeks to improve long-term risk-adjusted returns by making sustainability a core component of financial analysis.

ESG integration has become the fastest-growing segment of sustainable finance, now commanding the largest share of assets under management where environmental, social and governance factors are systematically applied. According to the latest figures, the global market for ESG-focused investments exceeded twenty-five trillion dollars (Global Sustainable Investment Alliance, 2023). This number might not be entirely accurate, but this surge reflects several converging trends:

1. Investors increasingly accept that ESG factors are financially material and vital for long-term performance.
2. Regulators have imposed mandates for climate risk disclosure and broader ESG reporting, as seen in the EU's SFDR and proposed SEC rules.
3. Data providers have improved both the availability and quality of ESG information despite standardization remains challenging.
4. A growing body of academic research links robust ESG integration to stronger financial results, lower capital costs and enhanced risk management.

A typical example would be the following case. An analyst covering the food industry integrates social factors by examining companies' supply chain labor practices. Company X sources ingredients from regions known for poor labor standards and lacks transparency regarding its suppliers, posing risks of operational disruptions, negative media campaigns and potential consumer boycotts.

As a contrast, Company Y invests heavily in supply chain auditing, fair wage initiatives and publicly reports on its supplier relationships. The analyst adjusts their financial model for Company X to account for higher potential costs and reduced revenues related to reputational damage and supply chain instability, while viewing Company Y's practices as a competitive advantage that supports brand loyalty and operational resilience, potentially justifying a higher valuation multiple. This case shows how ESG has become a widely accepted and used risk framework.

Impact Investing

Impact investing sets itself apart by intentionally seeking positive and measurable social and environmental benefits in tandem with financial returns. It focuses on areas where traditional markets often fall short such as extending renewable energy access to underserved regions, promoting sustainable agriculture, financing affordable housing, improving healthcare in low-income communities, advancing education technology and social tech (Spiess-Knafl & Scheck, 2023).

Unlike socially responsible investing, which aims to avoid harmful industries, or ESG integration, which incorporates material risks and opportunities into financial analysis, impact investing places the generation of demonstrable outcomes at its core. Investors define clear social or environmental objectives, accept returns that may range from below market to market rate and commit to rigorous measurement and transparent reporting of impact performance. This accountability ensures that capital is directed toward solutions with quantifiable benefits.

The impact investing market has grown rapidly, though it remains smaller than the broader ESG and SRI universe. The Global Impact Investing Network estimated assets under management at approximately 1.571 trillion US dollars managed by 3907 organizations in 2024. Growth has been strong, with surveys reporting a 21% percent compound annual growth rate over the five years leading up to 2024 (Hand et al., 2024).

This momentum is driven by rising awareness of global challenges, alignment with the United Nations Sustainable Development Goals (SDGs), supportive policy frameworks and growing investor demand for opportunities that deliver both financial return and positive societal change.

Examples for impact investing are fund who provide capital to a company providing off-grid solar systems to rural communities in Africa, extend debt to a microfinance institution serving women entrepreneurs or back a company focused on providing inclusive technology for persons with disabilities.

One example of an impact investing fund is the Munich-based fund BonVenture. It was established in 2003 and is one of the most experienced impact investment funds in Europe. At the beginning, it was commonly referred to as social venture capital or venture philanthropy (Heister, 2010). Over two decades, BonVenture has built a portfolio of companies, focusing primarily on Germany but also extending into other European countries (BonVenture, 2025).

The firm's evolution can be traced by the increasing fund sizes over time. The first fund was launched in 2003 with €5 million and the fourth fund had a volume of €50 million (Bruckschlögl, 2022). Other funds have been set up in Germany which follow a similar approach but with different investment strategies. Some are focusing more on climate tech, while others focus more on social impact (Freiburg & Spiess-Knafl, 2024) (Table 5.2).

Table 5.2 Categories of Sustainable Finance

Category	Impact Investing	SRI Investing	ESG Investing	Traditional Investing
Focus	Focus on Values	Focus on Values	Focus on Value	Focus on Value
Outcomes	Intentional focus on E or S outcomes	Usually driven by non-financial preferences	Some non-financial preferences, some value motivations	Usually value motivations
Return expectations	Some willing to accept lower returns for impact, but some expect risk-adjusted market returns	Expectations of closer to risk-adjusted market returns with some willing to accept lower returns	Expectation of risk-adjusted market returns or greater than risk-adjusted market returns, while some are willing to accept lower returns	Expectation of risk-adjusted market returns or greater than risk-adjusted market returns
Capital markets	Private capital markets	Public capital markets	Mostly public capital markets	Private and public capital markets
Market size	~ $1.57 Trillion (GIIN 2024 estimate)	Significant but often bundled; Exclusion strategy applied to tens of trillions globally	Largest segment within sustainable finance; ESG factors considered across > $25 trillion	Total global AUM (>$100 Trillion)
Examples	Microfinance funds, Renewable energy projects in developing nations, Affordable housing REITs	Excluding tobacco/weapons stocks, Sharia-compliant funds, Fossil-fuel-free portfolios	Analyzing carbon intensity for risk, Investing based on high ESG ratings (e.g., MSCI), Engaging on board diversity	S&P 500 index fund, Value stock investing based purely on financial metrics, Hedge funds

Source Own illustration based on Starks (2023) and market estimates based on Global Sustainable Investment Alliance (2023), and Hand et al. (2024)

5.4 Classification by Financial Structure

Sustainable finance instruments span the full range of traditional financial structures, each adapted to incorporate sustainability considerations. We can differentiate between three different instruments:

- Equity capital
- Debt capital
- Hybrid capital

Equity Instruments

Equity instruments give investors ownership stakes in companies, enabling them to pursue financial returns while supporting environmental and social goals. One common strategy in public equity markets is to integrate environmental, social and governance factors into investment analysis. Investors evaluate material ESG issues such as carbon emissions, board diversity, and labor practices to adjust company valuations and portfolio allocations in line with long-term risks and opportunities.

Another approach is thematic equity funds, which concentrate on companies in specific sustainability sectors such as renewable energy, sustainable agriculture or clean technologies. These funds seek to capture growth driven by global sustainability trends. In both integration and thematic strategies, active ownership is essential. Shareholders use voting and engagement to influence corporate behavior and improve sustainability practices from within (Dimson et al., 2015).

Equity investment as part of the impact investment field goes further by directing capital to companies whose core activities deliver measurable social or environmental benefits. That is common in private capital markets through venture capital and private equity vehicles funding mission driven enterprises.

Beyond impact-focused mandates, Environmental, Social and Governance (ESG) considerations are increasingly being integrated throughout the investment lifecycle in Private Equity (PE) and Venture Capital (VC). Unlike public market investors who primarily rely on engagement and voting, PE and VC firms leverage their significant ownership stakes and board representation to directly influence portfolio companies.

This allows ESG integration to move beyond basic screening toward driving operational improvements such as enhancing energy efficiency or resource management to cut costs and implementing stronger governance structures to build resilience and trust. The goal is often framed as using ESG as a cross-functional lever for value creation across the investment cycle.

Key drivers for this deeper integration include increasing pressure and specific mandates from Limited Partners (LPs), a desire to mitigate ESG-related risks and the recognition that strong ESG performance can enhance long-term value. This is often referred to as "multiple arbitrage" as sustainable companies might command higher exit proceeds. In addition, it helps to attract ethically minded consumers and future investors. This hands-on approach offers a different mechanism for ensuring capital supports sustainability goals compared to the thematic selection and shareholder engagement common in public equities. However, challenges remain, particularly in obtaining reliable and consistent ESG data from private portfolio companies and in standardizing methodologies for reporting and comparison (e.g., Clark & Louch, 2021). This is especially true in early-stage venture capital.

Equity differs from debt because it represents a claim on the entire enterprise rather than proceeds tied to projects. Green, social or sustainability bonds can restrict use of proceeds, but equity cannot. Instead, equity investors rely on thematic selection, impact focus and shareholder engagement to ensure capital supports sustainability goals.

There is thus no direct equity equivalent to sustainability-linked loans where financing terms adjust based on agreed ESG targets. However, equity holders still drive change by setting clear expectations through voting policies and ongoing dialogue.

Debt Instruments

Use of Proceeds vs. Performance-Linked Instruments

Debt instruments in sustainable finance can be grouped into two main categories. Use of proceeds instruments direct funds to designated projects, such as green bonds that finance renewable energy, social bonds that support community initiatives and sustainability bonds that back a combination of environmental and social programs. Sustainability-linked instruments, including sustainability-linked bonds and sustainability-linked loans, adjust financial terms based on the issuer's overall sustainability performance.

These instruments offer predictable income streams while steering capital toward clearly defined sustainability goals. Risk and return profiles vary with the credit quality of the issuer and the specific design features of each instrument, giving investors a range of options to align their fixed income investments with environmental and social objectives.

Green Bonds

Green bonds are fixed income instruments whose net proceeds support projects with clearly defined environmental benefits. Their purpose is to mobilize capital for initiatives such as large-scale solar and wind farms, energy-efficiency upgrades, sustainable water management and pollution prevention to help the transition to a low-carbon climate resilient economy and address other environmental challenges including resource depletion and biodiversity loss (Bhutta et al., 2022).

The credibility of green bonds is also shaped by frameworks that vary in scope and requirements. The table below compares the ICMA Green Bond Principles, the Climate Bonds Standard with its certification scheme and the EU Green Bond Standard (Table 5.3).

Green bonds feature the ring-fencing of proceeds, a transparent evaluation and selection process, dedicated tracking of funds and ongoing reporting to investors. In practice, issuers allocate those funds to a range of eligible projects with clear environmental benefits. They finance renewable energy developments such as large scale solar and wind farms, support energy-efficiency initiatives like building retrofits and smart grid installations. They also back projects such as sustainable water management schemes including wastewater treatment plants and flood protection. Green bonds also underwrite pollution prevention measures such as electrification of public transport networks and expanded recycling programs, as well as biodiversity conservation efforts like habitat restoration and sustainable forestry projects.

This overview shows the range of potential projects and it is also explains that the global green bond issuance exceeded $600 billion in 2024 growing from $270 billion in 2019 (Environmental Finance, 2025). The total amounts outstanding are around $3 trillion with the most bonds issued in Euro, followed by USD (Demski et al., 2025).

Table 5.3 Comparison of key green bond standards

Feature	ICMA Green Bond Principles (GBP)	Climate bonds standard & certification	EU Green Bond Standard (EuGBS)
Nature	Voluntary process guidelines	Voluntary labelling scheme	Voluntary label with mandatory requirements
Use of proceeds basis	Broad eligible project categories	Science-based sector criteria (Taxonomy)	EU Taxonomy regulation alignment
Taxonomy alignment	Recommended (e.g., EU Taxonomy)	Implicit via sector criteria	Mandatory (≥85% proceeds + DNSH)
External review	Recommended	Required for certification	Required (Pre- & Post-Issuance)
Reviewer oversight	Non-specific	CBI approved verifier network	ESMA registration & supervision
Reporting	Recommended (Annual allocation & impact)	Required (Annual)	Mandatory (Factsheet, allocation, impact)
Reporting format	Guidance (e.g., Harmonized framework)	Standard requirements	Mandated templates
Geographic scope	Global	Global	EU regulation (Globally applicable label)

Source Own illustration based on Climate Bonds Initiative (2023), European Parliament (2023), and International Capital Market Association (ICMA) (2022)

Europe is the frontrunner supported by the EU Sustainable Finance Framework, the EU Gren Bond Standard and the activities of supranational activities such as the European Investment Bank and the European Commission (European Environment Agency, 2024).

To strengthen market integrity issuers commonly obtain external reviews such as second party opinions or post-issuance verifications from providers like Sustainalytics and DNV to confirm alignment with the Green Bond Principles. While robust impact reporting is encouraged, standardizing methodologies and consistently quantifying environmental outcomes remains an ongoing area of market development. Furthermore, discussions around additionality underscore the importance of rigorous project selection and verification. Additionality is understood in this context as the extent to which bond funding enables projects that would not otherwise occur (Andersen et al., 2021).

A notable example of corporate green bond issuance is Apple, which has mobilized $4.7 billion since 2016 through three distinct offerings in both USD and EUR.

The first issuance, launched in 2016, marked the first green bond ever issued by a U.S. technology company and, at the time, the largest from any U.S. corporation. It had a 7-year maturity with a 2.85% coupon rate and was priced at a spread of 135 basis points over U.S. Treasuries (Volcovici, 2016). The second issuance followed in 2017 with comparable terms, raising $1 billion. In 2019, Apple entered the European market with a €2 billion issuance, structured in two tranches: €1 billion of 0.0% notes due 2025 and €1 billion of 0.5% notes due 2031.

Apple's program is structured around a comprehensive Green Bond Framework that is explicitly aligned with the four core components of the Green Bond Principles (GBP) discussed above. This framework underpins the company's approach to project evaluation, selection and the management of proceeds, ensuring transparency and credibility (Apple, 2024).

The proceeds from Apple's green bonds are exclusively directed toward projects with clearly defined environmental benefits, consistent with the use of proceeds nature of such instruments. Eligible categories span a wide range, including low-carbon design and engineering, energy-efficiency enhancements across corporate facilities and the manufacturing supply chain, substantial investments in renewable energy generation, carbon mitigation efforts, and carbon sequestration (Apple, 2022).

Social Bonds

Social bonds are fixed income instruments such as green bonds, but with proceeds ring-fenced for projects that generate positive social outcomes. Their primary purpose is to channel capital toward initiatives such as affordable housing, access to healthcare and education, employment generation, and socioeconomic empowerment. By dedicating funds to these areas, social bonds help reduce inequality, and improve community well-being. There is also first research about the existence of a social premium comparable to the green premium (Torricelli & Pellati, 2023).

The credibility of social bonds rests on the ICMA Social Bond Principles, which outline four core components: the exclusive use of proceeds for eligible social projects; a transparent process for project evaluation and selection; dedicated management and tracking of funds; and regular reporting on both allocation and social impact (International Capital Market Association (ICMA), 2023).

Typical eligible uses include the financing of affordable housing developments, construction or renovation of clinics and schools, microfinance and small business lending to create jobs and programs that enhance food security or support marginalized groups. These project categories align closely with United Nations Sustainable Development Goals such as No Poverty (SDG 1), Good Health and Well-Being (SDG 3), Quality Education (SDG 4), and Reduced Inequality (SDG 10).

The global social bond market has grown rapidly, surpassing USD 1.1 trillion in cumulative issuance by the end of 2024. Annual issuance spiked during the COVID-19 pandemic with a peak at over USD 200 billion in 2021, before normalizing to around USD 175 billion in 2022 and USD 181 billion in 2023 (Cochelin et al., 2024).

To reinforce market integrity, issuers commonly obtain external reviews such as Second Party Opinions or post-issuance verifications from providers like Sustainalytics and DNV. While robust impact reporting is encouraged, standardizing social outcome metrics and demonstrating additionality remain critical areas of ongoing market development. This remains true whether financed projects would have occurred without bond proceeds.

A good example is Danone's issuance of a social bond in March 2018 with a volume of €300 million, a 1% coupon and a duration of 7 years. The bond was aligned with the International Capital Market Association (ICMA) Social Bond Principles (Environmental Finance, 2019).

The proceeds were allocated across five key areas: Responsible Farming and Agriculture, Social Inclusiveness, Research & Innovation for Advanced Medical Nutrition, Entrepreneurship Financing, and Quality Healthcare and Parental Support. The largest amount of €122 million went to the Utrecht R&I Center for Advanced Medical Nutrition, another €49 million was allocated to the Danone Ecosystem Fund and €45 million were allocated to the Danone US Pledge of not using GMO ingredients (Danone, 2019).

Sustainability Bonds

Sustainability bonds are debt instruments whose net proceeds are exclusively applied to finance or refinance a combination of both environmental and social projects. Unlike green bonds (which focus solely on environmental outcomes) or social bonds (which focus solely on social outcomes), sustainability bonds require issuers to fund initiatives across both domains. Their purpose is to provide a single financing tool that channels capital toward a broad range of sustainability objectives such as renewable energy and energy efficiency alongside affordable housing and access to health and education services (Mocanu et al., 2021).

Sustainability bonds adhere to the International Capital Market Association's Sustainability Bond Guidelines, which build on the Green Bond Principles and the Social Bond Principles (International Capital Market Association (ICMA), 2021). These voluntary guidelines prescribe four core components:

1. **Use of Proceeds**: Net proceeds must be allocated to a portfolio combining eligible "green" and "social" projects. Issuers must clearly define both categories and the types of projects financed.
2. **Process for Project Evaluation and Selection**: Issuers must disclose their sustainability objectives, eligibility criteria and selection process for both environmental and social projects, as well as any exclusion policies and risk-management procedures.
3. **Management of Proceeds**: Proceeds must be credited to a dedicated account or sub-portfolio and tracked through internal systems to ensure full allocation to the defined projects.
4. **Reporting**: Issuers should publish at least annual reports until proceeds are fully allocated, detailing the projects financed and providing narrative or quantitative measures of expected and actual environmental and social impacts.

Projects eligible under sustainability bonds span the full spectrum of ICMA's green and social categories. On the environmental side, this includes renewable energy production, energy-efficiency upgrades in buildings and industry, sustainable water and wastewater management, pollution-control technologies, biodiversity conservation, climate-adaptation infrastructure and circular-economy innovations. On the social side, eligible projects cover affordable basic infrastructure (clean water, sanitation, transport), access to essential services (healthcare, education, financial inclusion), affordable housing construction and renovation, employment-generation programs (including SME financing and microfinance), food-security initiatives and socioeconomic empowerment efforts targeting vulnerable populations. By financing these diverse project types under one instrument, sustainability bonds allow issuers to address interconnected sustainability challenges and offer investors a broad, impact-oriented exposure.

The global market for Green, Social, Sustainability and Sustainability-Linked (GSSS) bonds surpassed USD 1 trillion in annual issuance in 2024. Within this universe, sustainability bonds occupy a distinct niche which is a bit smaller than the social bonds. Cumulative aligned sustainability bond volume approached USD 1 trillion by late 2024, led primarily by development banks, supranational and sovereign issuers capable of managing the dual-objective mandate (Cochelin et al., 2024).

To reinforce market integrity, issuers commonly obtain independent external reviews such as Second Party Opinions or post-issuance verifications. These reviews assess alignment with ICMA's core components and help guard against "impact washing."

While robust impact reporting is encouraged by the guidelines, standardizing methodologies and consistently quantifying combined environmental and social outcomes remains an evolving area of market practice.

One example is the African Development Bank. Its sustainable bond program updates and combines its previous green bond programs and social bond programs. It can use the proceeds to take measures against climate change and support societies in the member countries (African Development Bank Group, 2023).

In 2024, the bank launched the first sustainable hybrid capital transaction from a multilateral development bank. The USD 750 million perpetual subordinated notes were issued under the Bank's 2023 Sustainable Bond Framework and designed to strengthen its capital base while expanding lending capacity for green and social projects. The hybrid instrument features a fixed annual coupon of 5.75% until August 2034, after which it resets every five years. The demand for the bonds peaked at USD 6 billion from more than 275 investors, with a majority coming from the UK, followed by Europe, Asia and the Americas (African Development Bank, 2024).

Sustainability-Linked Loans and Bonds

Sustainability-linked loans and sustainability-linked bonds tie financing costs to an organization's achievement of predefined sustainability performance targets, rather than earmarking proceeds for specific projects. This allows funds to be used for general corporate or sovereign purposes and offers flexibility for entities in transition-intensive sectors or governments pursuing broad policy goals. By adjusting loan spreads or bond coupons based on improvements in measures such as greenhouse gas emissions reductions or leadership diversity, these instruments create a direct financial incentive for continuous environmental, social and governance progress (Auzepy et al., 2023).

Market participants adhere to voluntary industry guidelines to safeguard credibility. For example, the bond market relies on the Sustainability-Linked Bond Principles published by the International Capital Market Association, which require public annual reporting and external verification as well as clear bond terms that specify how coupon adjustments reflect target achievement (International Capital Market Association (ICMA), 2024).

Common performance indicators suggested in these standards include reductions in Scope 1 and Scope 2 emissions, increases in renewable energy use, energy intensity improvements and enhanced water efficiency. Social metrics often focus on greater gender diversity in leadership, workplace safety and expanded access to essential services. Each target must represent a material improvement over a defined baseline and where possible align with science-based pathways or national objectives.

Sustainability-linked loans accounted for 650 billion euros in 2023 which represented 72% of total sustainable loan volume according to numbers from (BBVA, 2025). By contrast, sustainability-linked bond issuance fell from its 2021 peak to about 66 billion dollars in 2023 and 38 billion dollars in 2024 (Cochelin et al., 2024).

An illustrative example can be found in Italy. The utility company Enel is active in 28 countries and the largest private renewable energy operator as well as the largest private distribution company. In total, they serve 54.8 million households and businesses. The company aims to reduce all emissions and has a 2040 Net Zero decarbonization strategy with the following targets (Enel, 2025b):

- New renewable capacity deployed with a 100% GHG-free fleet by 2040;
- The exit from coal-based generation by 2027 and from gas-based generation by 2040;
- The exit from the gas retail business by 2040 and the achievement of 100% electricity sold to customers from GHG-free sources by 2040;

- The rollout of a capex plan in full alignment with the 2040 Zero Targets.

These targets are expensive but they are also clearly related to a reduction in greenhouse gas emissions. For this reason, the company is a pioneering issuer in the sustainability-linked bond (SLB) market, having launched the first-ever SLB in September 2019 with a volume of $1.5 billion and an effective yield of 2.676% (Enel, 2019). Over time, the company issue with multiple sustainability-linked bonds following its own framework. The most recent issuance had 3 tranches and one tranche had the following structure (Enel, 2025a).

Structure: 750 million euros at a fixed rate of 2.625%, with settlement date set on February 24, 2025, maturing February 24, 2028:

- the issue price has been set at 99.574% and the effective yield at maturity is equal to 2.775%;
- the interest rate will remain unchanged to maturity, subject to the joint achievement of the following Sustainability Performance Targets ("SPTs"). In particular:
- for the KPI related to the "Proportion of CAPEX aligned to the EU Taxonomy (%)," the achievement of a SPT equal to or higher than 80% on December 31st, 2025 for the 2023–2025 period;
- for the KPI related to the "Scope 1 GHG emissions Intensity relating to Power Generation (gCO_{2eq}/kWh)," the achievement of a SPT equal to or less than 130gCO_{2eq}/kWh on December 31, 2025;
- if one or both of the abovementioned SPTs are not achieved, a step-up mechanism will be applied, increasing the rate by 25 bps, as of the first interest period subsequent to the publication of the relevant assurance report issued by an external verifier;

Blended Finance and the Impact Enterprise Funding Ecosystem

Impact enterprises often rely on a complex ecosystem of funding and support to achieve both financial sustainability and positive social or environmental outcomes. As depicted in Fig. 5.1, this ecosystem comprises three layers: (1) Capital Structure, (2) Operating Structure, and (3) Revenue Structure. Each layer engages different actors and financing mechanisms.

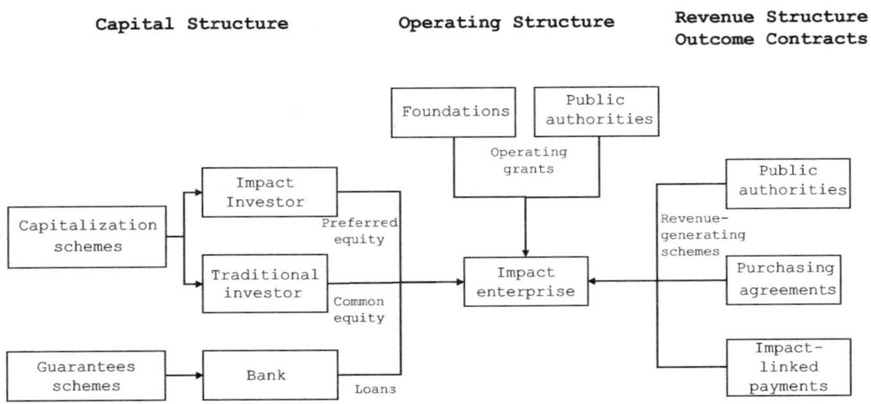

Fig. 5.1 Layers in hybrid finance. *Source* Own illustration

Blended finance thus refers to the strategic use of public or philanthropic capital to mobilize private investment in sustainable development projects, particularly in high-risk or emerging markets where traditional investors may be reluctant to commit resources. By deploying grants, concessional loans or guarantees, blended finance bridges funding gaps, enhances project viability and mobilizes the substantial private capital to achieve predefined social outcomes (Vecchi et al., 2017).

Within the Capital Structure layer, impact enterprises draw on a mix of equity and debt. Impact-focused investors may provide preferred or common equity in pursuit of both financial returns and measurable impact, while traditional investors supply common equity primarily seeking market-rate returns. Public and philanthropic actors can play a catalytic role by anchoring funds with large equity investments in the range of ten to twenty million euros. The catalytic effect arises through the de-risking of the investment opportunity and thus attracting follow-on private financing. On the debt side, banks and other lenders extend loans, often supported by guarantee schemes that mitigate credit risk (Spiess-Knafl & Scheck, 2023). The European Investment Fund, for example has been found to be instrumental in unlocking billions for micro-enterprises and social enterprises (Haarich et al., 2020). Development finance institutions also provide debt on concessional terms, further strengthening the financing base available to impact ventures.

Parallel to investment capital, the Operating Structure layer supplies grant funding that underpins core operations and early-stage development. Europe alone is home to roughly 186,000 foundations distributing an estimated 54.5 billion euros annually, with most grants averaging around 300,000 euros each and earmarked for specific projects (e.g., Musaddique, 2024). National and local governments contribute still larger sums through social budgets and dedicated programs; for instance, Germany allocates approximately 26.7% of its GDP to social expenditures, a portion of which supports impact enterprises delivering essential services. While grants are vital for covering non-commercial activities and absorbing initial risks, their contractual restrictions such as prohibitions on double funding and limits on covering financing costs require careful coordination with investment capital to preserve entrepreneurial agility (Achleitner et al., 2014).

The third layer with the revenue structure is based on performance-linked contracts to generate sustainable income. Public authorities may enter into service agreements that pay the enterprise for delivering specified outcomes or achieving operational milestones. Corporations and government agencies can establish pre-purchase arrangements, guaranteeing revenue for the goods or services produced. A recent example were pre-purchase agreement for carbon capturing technologies is Frontier which was founded and funded by Stripe, Alphabet, Shopify, Meta and McKinsey (Frontier, 2022).

In addition, impact-linked payment models tie remuneration directly to the achievement of agreed social or environmental targets, aligning cash flow with mission delivery. These outcome-based contracts transform the enterprise's core impact activities into predictable revenue streams, reducing dependency on grants or investor funding alone.

The interplay of these three layers offers unparalleled flexibility in the impact finance ecosystem. Foundations can combine grant awards with guarantees or anchor investments. Corporations may support enterprises through equity stakes, advance purchase agreements or outcome-linked contracts. Public authorities provide a mix of grants, catalytic capital, and service-based payments. Wealthy individuals, pension funds and development banks similarly tailor their engagement across grants, concessional loans, equity, and outcome-linked financing. By accommodating diverse preferences for risk, return, and impact, this layered blended finance model creates a robust system capable of channeling both concessional and commercial capital toward enterprises that generate social and environmental value.

Market-Based Mechanisms

Instruments in this category include carbon markets (both compliance and voluntary), payments for ecosystem services, biodiversity credit, and token-based impact models among others. These instruments create financial value for environmental assets previously considered externalities, enabling market-based solutions to environmental challenges. There has been significant amounts of experimentation with blockchain-based tokens in the past years (Spiess-Knafl, 2022).

Carbon markets facilitate the trading of carbon credits, enabling companies to offset their greenhouse gas emissions by investing in projects that reduce emissions elsewhere. These markets are divided into compliance markets, which are regulated by government bodies such as the European Union Emissions Trading System and voluntary markets, where trading is driven by corporate sustainability commitments and broader social responsibility initiatives. By assigning a price to carbon dioxide, carbon markets provide a financial incentive for companies to reduce emissions and invest in cleaner technologies, supporting the overall goal of lowering global greenhouse gas levels.

According to the World Bank's (2024) report on "State and Trends of Carbon Pricing," carbon pricing instruments currently cover around 24% of global greenhouse gas emissions, though there remains a substantial implementation gap between national commitments and actual policy measures. The report notes that 75 carbon taxes and emissions trading systems are in operation worldwide, with some instruments showing modest gains while others have experienced price decreases over the past year. Despite this growth, price levels in many established systems continue to fall short of what is needed to meet the Paris Agreement goals.

Revenues from carbon pricing reached record highs in 2023, exceeding USD 100 billion, with emissions trading systems accounting for most of these revenues. However, even with these high revenue figures, the contribution of carbon pricing to national budgets remains relatively small. This indicates both the potential and the limitations of current pricing policies as fiscal tools (World Bank, 2024).

5.5 The Bottom Line: Financial Returns of Sustainable Finance

The ESG Performance Puzzle

Sustainable finance, which integrates environmental, social, and governance criteria into financial decisions, has moved from a niche topic to a central issue in global markets. Assets managed according to sustainable principles have grown rapidly as outlined above. Investors, regulators and researchers alike ask whether applying ESG criteria leads to different financial returns than conventional strategies. This question arises because ESG issues introduce risks and opportunities not always captured in traditional analyses and because investor motivations range from financial objectives to mixed financial and social preferences.

Evaluating the performance of ESG investing is challenging. Different rating agencies often assign very different ESG scores to the same company, owing to variations in methodology, data sources and weighting. This rating divergence makes it hard to know which scores to trust when constructing portfolios. Moreover, studies mix many approaches, from values driven strategies that prioritize social goals to value driven strategies that use ESG mainly for risk management. To understand return implications, one must untangle these varied strategies and data issues.

The Link Between ESG and Corporate Financial Performance

A large body of research has examined how a company's own ESG activities affect its financial performance, separate from how ESG portfolios perform. The most comprehensive meta-analysis, covering roughly 2,200 empirical studies since the 1970s, finds that about 90% report a non-negative relation between ESG and firm performance, with most showing positive effects. Integrating sustainability into business practices seldom imposes a financial cost and often brings benefits that hold up over time (Friede et al., 2015).

The channels driving this positive link include higher employee motivation and retention, stronger brand reputation and stakeholder relationships, more efficient investment and growth, greater capacity for dividend payments, lower systematic and idiosyncratic risks, and reduced costs of debt and equity alongside higher optimal leverage. While some studies report mixed results on direct profitability metrics such as return on assets or point to potential agency costs of certain CSR activities, the overall evidence indicates that ESG performance enhances firm value through multiple reinforcing mechanisms.

ESG Investment Portfolio Returns

Although strong ESG performance at the corporate level often translates into better firm results, the picture is more nuanced for investment portfolios. On average, ESG portfolios such as mutual funds or indices do not underperform conventional benchmarks and may deliver neutral to slightly positive returns on a risk-adjusted basis. The meta-analysis by Friede and others (2015) finds that only around 15% of portfolio based studies report positive results, compared with about 57% of non-portfolio studies. This suggests that while ESG integration benefits individual firms, achieving consistent portfolio alpha after costs and diversification is more difficult.

Still, some recent analyses document significant excess returns. Berg et al. (2023) show that in the United States and Japan, top versus bottom quartile portfolios based on certain ESG scores generated statistically significant positive alphas over factor models from 2014 to 2020. These gains varied widely depending on which rating agency data was used, implying that rating choice may drive a part of the return differential. Other studies find ESG portfolios suffer smaller drawdowns in market crises, hinting at downside protection. In the bond market, green bonds were found to yield no material difference from conventional bonds (Zerbib, 2019).

Explaining the Variance: Factors Influencing Returns

Several factors help explain why studies reach different conclusions about ESG portfolio returns (Berg et al., 2022; Friede et al., 2015; Pedersen et al., 2021):

- ESG data quality and divergence among providers create noise that makes it difficult to isolate a clear ESG factor and may affect returns depending on which ratings one uses
- Aggregation methods that combine multiple ratings can reduce noise and enhance the ESG signal, often producing stronger alphas than single source ratings.
- Portfolio versus non-portfolio studies yield different findings because diversified funds face trading and management costs and may have sector or style exposures that dilute firm-level ESG benefits.
- Time period and market dynamics matter, with earlier studies offering mixed results due to poor data and simple strategies, while more recent periods show pockets of positive excess returns linked to higher ESG awareness and fund flows and resilience in crises such as COVID-19.
- Investor preferences also matter, as some investors accept lower expected returns in exchange for social or environmental impact, creating equilibrium effects in which green assets may trade at a slight premium.

Strong evidence supports the idea that good ESG performance enhances firm-level results but translating that into superior portfolio returns is not happening automatically. Recent work finds episodes of positive excess returns for ESG strategies, likely driven by capital flows and evolving investor tastes, but outcomes depend heavily on data source, aggregation approach, market period, and regional context. The key challenge remains rating divergence, though combining multiple data sources shows promise. Ultimately, evaluating ESG returns requires distinguishing firm value creation from portfolio outcomes while recognizing the roles of data quality, methodology, market conditions and non-financial investor motivations.

5.6 Fundamentals of the Return on Sustainable Finance

Valuing a sustainable finance investment follows the same basic logic as any other investment opportunity. It is done by estimating the present value of the cash flows it will generate over time. In practice, that means using a discounted cash flow model to bring future revenues and costs back to today's terms.

What makes sustainable finance distinct, however, is the need to incorporate additional dimensions alongside plain financial performance. For example, a wind farm or a solar park might benefit from higher feed-in tariffs or carbon credits that can be sold. A consumer-facing ethical brand might have more loyal customers which might also have a higher willingness to pay (Cachero-Martínez et al., 2023).

Central to any discounted cash flow analysis is the choice of discount rate. In impact investing, that rate is often considerably lower than a strictly market-based assessment would suggest. The reason is not that renewable energy projects or affordable housing funds are somehow less risky in a traditional sense, but that the ultimate investors which are often referred to as LPs (limited partners) accept below market returns in exchange for measurable non-financial outcomes. By applying a lower discount rate in their models, impact managers build into their valuations the mandate to deliver quantifiable carbon reductions, improved livelihoods, or other social gains alongside modest internal rates of return. This approach allows funds to proceed with projects that might not meet conventional return thresholds, but nonetheless generate the kind of long-term value that aligns with both investor intentions and broader sustainability goals.

References

Achleitner, A.-K., Spiess-Knafl, W., & Volk, S. (2014). The financing structure of social enterprises: Conflicts and implications. *International Journal of Entrepreneurial Venturing, 6*(1), 85–99.

African Development Bank. (2024). *African Development Bank launches the first ever Sustainable Hybrid Capital transaction from an MDB: a USD 750 million PerpNC10.5 Sustainable Benchmark Hybrid.* https://www.afdb.org/sites/default/files/documents/afdb_inaugural_sustainable_hybrid_usd_750m_perpnc10.5_benchmark_transaction_-_press_release.pdf

African Development Bank Group. (2023). *Sustainable bond program.* African Development Bank Group. https://www.afdb.org/en/financial-information/investor-resources/capital-markets/sustainable-bond-program

Andersen, O. W., Hansen, H., & Rand, J. (2021). Evaluating financial and development additionality in blended finance operations. *OECD Development Co-Operation Working Papers.* https://www.oecd.org/content/dam/oecd/en/publications/reports/2021/02/evaluating-financial-and-development-additionality-in-blended-finance-operations_9940d03d/a13bf17d-en.pdf

Apple. (2022). *Apple's $4.7B in Green Bonds support innovative green technology.* https://www.apple.com/newsroom/2022/03/apples-four-point-seven-billion-in-green-bonds-support-innovative-green-technology/

Apple. (2024). *Annual Green Bond impact report.* https://s2.q4cdn.com/470004039/files/doc_downloads/additional_reports/2023/apple_greenbond_report_fy2023.pdf

Auzepy, A., Bannier, C. E., & Martin, F. (2023). Are sustainability-linked loans designed to effectively incentivize corporate sustainability? A framework for review. *Financial Management, 52*(4), 643–675. https://doi.org/10.1111/fima.12437

BBVA. (2025). *Green and sustainability-linked loan.* BBVA CIB. https://www.bbvacib.com/green-and-sustainability-linked-loan-newsletter/

Berg, F., Kölbel, J. F., & Rigobon, R. (2022). Aggregate confusion: The divergence of ESG ratings. *Review of Finance, 26*(6), 1315–1344.

Berg, F., Lo, A. W., Rigobon, R., Singh, M., & Zhang, R. (2023). *Quantifying the returns of ESG investing: An empirical analysis with six ESG metrics.* https://papers.ssrn.com/sol3/papers.cfm?abstract_id=4367367

Berry, T. C., & Junkus, J. C. (2013). Socially responsible investing: An investor perspective. *Journal of Business Ethics, 112*(4), 707–720. https://doi.org/10.1007/s10551-012-1567-0

Bhutta, U. S., Tariq, A., Farrukh, M., Raza, A., & Iqbal, M. K. (2022). Green bonds for sustainable development: Review of literature on development and impact of green bonds. *Technological Forecasting and Social Change, 175*, Article 121378. https://doi.org/10.1016/j.techfore.2021.121378

BonVenture. (2025). *Impact startup portfolio.* https://www.bonventure.de/portfolio/

Bruckschlögl, R. (2022). *Final closing: Bonventure raises 50 million euros.* https://www.munich-startup.de/en/80922/final-closing-bonventure-fund-iv/

Cachero-Martínez, S., García-Rodríguez, N., & Salido-Andrés, N. (2023). Because I'm happy: Exploring the happiness of shopping in social enterprises and its effect on customer satisfaction and loyalty. *Management Decision, 62*(2), 492–512. https://doi.org/10.1108/MD-11-2022-1536

Clark, S., & Louch, W. (2021). *The key man: The true story of how the global elite was duped by a Capitalist Fairy Tale.* Harper Business.

Climate Bonds Initiative. (2023, May 21). *The climate bonds standard.* Climate Bonds Initiative. https://www.climatebonds.net/standard/the-standard

Cochelin, P., Popoola, B., & Volland, E. (2024). *Sustainable bond issuance to approach $1 Trillion in 2024.* https://www.spglobal.com/_assets/documents/ratings/research/101593071.pdf

Danone. (2019). *Social bond report.* https://www.danone.com/content/dam/corp/global/danonecom/investors/en-social-bond/2018/socialbond/2019-social-bond-report.pdf

Demski, J., Dong, Y., McGuire, P., & Mojon, B. (2025). Growth of the green bond market and greenhouse gas emissions. *BIS Quarterly Review.* March. https://www.bis.org/publ/qtrpdf/r_qt2503d.htm

Dimson, E., Karakaş, O., & Li, X. (2015). Active ownership. *The Review of Financial Studies, 28*(12), 3225–3268. https://doi.org/10.1093/rfs/hhv044

Eccles, R. G., & Klimenko, S. (2019). Shareholders are getting serious about sustainability. *Harvard Business Review, 97*(3), 106–116.

Enel. (2019). *Enel Launches the World's First General Purpose SDG-Linked Bond.* https://www.enel.com/content/dam/enel-common/press/en/2019-September/SDG%20bond%20ENG%20(003).pdf

Enel. (2025a). *Enel successfully launches a triple-tranche 2 billion euro "Sustainability-linked bond" in the Eurobond market.* https://www.enel.com/media/explore/search-press-releases/press/2025/02/enel-successfully-launches-a-triple-tranche-2-billion-euro-sustainability-linked-bond-in-the-eurobond-market

Enel. (2025b). *Sustainability-linked finance.* https://www.enel.com/investors/investing/sustainable-finance/sustainability-linked-finance

Environmental Finance. (2019). *Social bond of the year: Danone.* https://www.environmental-finance.com/content/awards/green-social-and-sustainability-bond-awards-2019/winners/social-bond-of-the-year-corporate-danone.html

Environmental Finance. (2025). *Resilience, innovation and reinvention: The sustainable bond market in 2025*. Environmental Finance. https://www.environmental-finance.com/content/the-green-bond-hub/resilience-innovation-and-reinvention-the-sustainable-bond-market-in-2025.html

European Environment Agency. (2024, November 4). *Green bonds in Europe*. https://www.eea.europa.eu/en/analysis/indicators/green-bonds-8th-eap

European Parliament. (2023). *Regulation (EU) 2023/2631 of the European Parliament and of the Council of 22 November 2023 on European Green Bonds and optional disclosures for bonds marketed as environmentally sustainable and for sustainability-linked bonds*. http://data.europa.eu/eli/reg/2023/2631/oj/eng

Freiburg, M., & Spiess-Knafl, W. (2024). Finanzierung von gemeinwohlorientierten Unternehmen in Deutschland mit Impact Investing. In *Social Entrepreneurship in Deutschland* (pp. 268–281). transcript Verlag. https://www.utb.de/doi/abs/10.36198/9783838563336-268-281

Friede, G., Busch, T., & Bassen, A. (2015). ESG and financial performance: Aggregated evidence from more than 2000 empirical studies. *Journal of Sustainable Finance & Investment*, *5*(4), 210–233.

Frontier. (2022). *An advance market commitment to accelerate carbon removal*. Frontier. https://frontierclimate.com/

Global Sustainable Investment Alliance. (2023). *Global sustainable investment review 2022*. https://www.gsi-alliance.org/members-resources/gsir2022/

Haarich, S., Holstein, F., Spule, S., Galera, G., Franchini, B., Borzaga, C., Chiomento, S., Scheck, B., Spiess-Knafl, W., & Salvatori, G. (2020). *Impact of the European commission's Social Business Initiative (SBI) and its follow-up actions*. Study for DG Employment, Social Affairs and Inclusion, European Commission.

Hand, D., Ulanow, M., Pan, H., & Xiao, K. (2024). *Sizing the impact investing market 2024*. The GIIN. https://thegiin.org/publication/research/sizing-the-impact-investing-market-2024/

Heister, P. (2010). *Finanzierung von social entrepreneurship durch venture philanthropy und social venture capital*. Springer.

International Capital Market Association (ICMA). (2021). *Sustainability bond guidelines*. https://www.icmagroup.org/sustainable-finance/the-principles-guidelines-and-handbooks/sustainability-bond-guidelines-sbg/

International Capital Market Association (ICMA). (2022). *Green bond principles*. https://www.icmagroup.org/sustainable-finance/the-principles-guidelines-and-handbooks/green-bond-principles-gbp/

International Capital Market Association (ICMA). (2023). *Social Bond Principles (SBP) ICMA*. https://www.icmagroup.org/sustainable-finance/the-principles-guidelines-and-handbooks/social-bond-principles-sbp

International Capital Market Association (ICMA). (2024). *Sustainability-Linked Bond Principles (SLBP)*. https://www.icmagroup.org/sustainable-finance/the-principles-guidelines-and-handbooks/sustainability-linked-bond-principles-slbp

Mocanu, M., Constantin, L.-G., & Cernat-Gruici, B. (2021). Sustainability bonds. An international event study. *Journal of Business Economics and Management*, *22*(6), Article 6. https://doi.org/10.3846/jbem.2021.15372

Musaddique, S. (2024, January 30). Europe home to over 180,000 public benefit foundations worth €647.5 billion. *Alliance Magazine*. https://www.alliancemagazine.org/blog/european-home-to-over-180000-public-benefit-foundations-worth-e647-5-billion/

Pedersen, L. H., Fitzgibbons, S., & Pomorski, L. (2021). Responsible investing: The ESG-efficient frontier. *Journal of Financial Economics, 142*(2), 572–597. https://doi.org/10.1016/j.jfineco.2020.11.001

Social Finance. (2017). *Reducing reoffending in Peterborough*. Social Finance. https://www.socialfinance.org.uk/work/reducing-reoffending-in-peterborough

Spiess-Knafl, W. (2022). *Artificial Intelligence and Blockchain for social impact: Social business models and impact finance*. Taylor & Francis Ltd.

Spiess-Knafl, W., & Scheck, B. (2023). *Impact investing: Instruments, mechanisms and actors* (2nd ed.). Palgrave Macmillan.

Starks, L. T. (2023). Presidential address: Sustainable finance and ESG Issues—Value versus values. *The Journal of Finance, 78*(4), 1837–1872. https://doi.org/10.1111/jofi.13255

Torricelli, C., & Pellati, E. (2023). Social bonds and the "social premium." *Journal of Economics and Finance, 47*(3), 600–619. https://doi.org/10.1007/s12197-023-09620-3

Vecchi, V., Balbo, L., Brusoni, M., & Caselli, S. (2017). *Principles and practice of impact investing: A catalytic revolution*. Routledge.

Volcovici, V. (2016). *Apple issues $1.5 billion in green bonds in first sale*. Reuters. https://www.reuters.com/article/technology/apple-issues-15-billion-in-green-bonds-in-first-sale-idUSKCN0VQ2K2/

World Bank. (2024). *State and trends of carbon pricing 2024*. https://openknowledge.worldbank.org/entities/publication/b0d66765-299c-4fb8-921f-61f6bb979087

Zerbib, O. D. (2019). The effect of pro-environmental preferences on bond prices: Evidence from green bonds. *Journal of Banking & Finance, 98*, 39–60. https://doi.org/10.1016/j.jbankfin.2018.10.012

CHAPTER 6

The Data Foundation of Sustainable Finance

Abstract This chapter examines the data foundation that is central for the Sustainable Finance Operating System, highlighting the indispensable role of reliable information while confronting the significant challenges posed by ESG data complexity. It applies an information theory lens to understand problems like signal to noise ratios and informational entropy in ESG datasets, underscoring the implications of this noisy environment for decision-making. The chapter also explores strategies for managing this complexity through technological advancements like artificial intelligence and the availability of satellite data, alongside the push for global standardization in reporting and taxonomies. Furthermore, the chapter maps the ESG data value chain, detailing the stages from production and verification to distribution, interpretation and activation. It discusses the specific technology solutions, including data-collection tools, processing systems, advanced analytics and visualization platforms. The role of satellite data as an independent verification source and the function of semantic layers in harmonizing disparate information are explored. Finally, the chapter analyzes the ESG ratings ecosystem, covering the purpose and construction of ratings, the landscape of providers, their market influence and the persistent challenges related to rating divergence, transparency and materiality definitions.

W. Spiess-Knafl, *The Sustainable Finance Operating System*, Palgrave Studies in Impact Finance,
https://doi.org/10.1007/978-3-031-97499-1_6

Keywords ESG data complexity · Satellite data · Data value chain · ESG ratings

6.1 The Role of Data in Sustainable Finance

Sustainable finance aims to reshape global capital flows, but this transformation is not feasible without reliable and actionable data. In the Sustainable Finance Operating System, ESG information is powering every decision from risk assessment to portfolio allocation. This operating system encompasses the network of regulations, disclosure standards, financial instruments and data platforms that together enable capital to be directed toward environmental and social objectives. Over time, sustainability reporting has evolved from basic carbon metrics into a multi-dimensional ecosystem drawing on corporate disclosures, regulatory filings, third-party ratings and even satellite imagery. It continues to evolve rapidly.

This data imperative faces a core challenge: ESG information remains far less standardized than traditional financial data. Methodologies vary widely among providers, definitions diverge on what constitutes an environmental or social metric and gaps persist where data is simply unavailable (Adams & Abhayawansa, 2022). It is also understandable considering the difficulties in assessing and measuring indicactors such as social equality, biodiversity, quality of life, or higher education (e.g., Hazelkorn & Gibson, 2019).

The result is an inconsistent patchwork that hinders meaningful comparisons and undermines confidence in ESG scores as reliable inputs for investment and lending decisions.

This lack of standardization leads directly to market inefficiencies and elevates risk. Investors and lenders operating on flawed or incomplete ESG data confront information asymmetry, which drives mispricing of risk and misallocation of capital as it can be observed across markets (Akerlof, 1978). In the absence of clear, comparable metrics, light forms of greenwashing can flourish, as issuers exploit reporting gaps to overstate their sustainability credentials. Such outcomes strike at the heart of sustainable finance's purpose, preventing markets from accurately rewarding genuine sustainability performance and impeding progress toward global goals.

To address these challenges, the Sustainable Finance OS relies on robust technical architecture and advanced tools. The data value chain extends from raw data collection through cleansing, integration, and analysis to the generation of decision-ready insights. Semantic layers organize disparate data sources into a coherent whole, while artificial intelligence helps to automate pattern recognition and anomaly detection. Together, these technologies can lift data quality, expand coverage and bridge the gap between complex sustainability impacts and quantitative analysis.

Mastering ESG data is a strategic imperative for companies, investors, and regulators alike. High-quality sustainability information underpins effective risk management, points toward new opportunities, ensures compliance with evolving frameworks and delivers competitive advantage. The rest of this chapter will map the sustainable finance data value chain, examine the technical architectures and tools that support it, explore the role of semantic layers in connecting information across standards and unpack the process by which ESG ratings and metrics are produced, validated and applied.

6.2 Understanding ESG Data Complexity: An Information Theory Perspective

The Challenge: Navigating ESG Data Complexity

Unlike traditional financial data where values like revenue, profit, and share price are governed by clear accounting standards ESG data comes in many forms. In addition, it is drawn from disparate sources and governed by varied expectations. This fundamental difference creates a level of complexity that must be managed before ESG data can reliably inform investment and policy decisions.

First, the sheer volume of ESG data is staggering. Companies report hundreds of metrics across environmental, social and governance dimensions. Single news aggregators provide access to more than 50,000 newspapers, magazines and websites. Third-party providers generate additional scores and benchmarks. Regulators demand ever more detailed disclosures. Tens of thousands of news media organizations report daily about relevant activities. Satellite data providers add continuous streams of environmental observations. Taken together, the potential universe of data points stretches into the millions for any given portfolio, far exceeding the relatively finite set of financial indicators.

Second, the variety and heterogeneity of ESG data amplify the challenge. Structured numbers in sustainability reports sit alongside narrative descriptions of labor practices. Media coverage and social media commentary introduce unstructured text which needs to be understood in context. Ratings and rankings from different agencies offer yet another perspective, each with its own methodology. Satellite imagery and Internet of Things sensors add geographic and real time dimensions. Reconciling such a mix of formats and sources into a coherent dataset requires substantial effort.

Third, veracity and quality concerns compound the problem. Much ESG information is self-reported by companies, raising questions about bias and completeness. Third-party estimates and proxies fill gaps but introduce their own uncertainties. Verification is often limited in scope and data may be inconsistent across reporting periods or incomparable across frameworks.

This combination of massive volume, extreme variety and variable quality creates a risk of information overload. In addition, market participants face a deluge of signals, only some of which are meaningful for assessing risk and opportunity. Sifting through this data to extract clear, material insights are at the heart of the sustainable finance challenge.

This inherent complexity is not merely a technical hurdle. It has profound consequences for the speed, accuracy and confidence with which sustainable finance decisions can be made.

Implications: Decision-Making in a Noisy Environment

The complexity of ESG data described in the previous section manifests directly in the day-to-day decisions of market participants. What begins as an abundance of potentially valuable information quickly becomes a stream of competing metrics, estimates, and narratives that obscure rather than illuminate true sustainability performance.

On one hand, the richness of multifaceted ESG data promises deeper nuance in assessing risk and opportunity. On the other hand, the sheer volume and variety can overwhelm even the most sophisticated analysis tools. Users confront information overload as they struggle to absorb streams of self-reported metrics, third-party scores, news feeds and sensor readings. In this environment, important signals become masked by noise, increasing the likelihood of mispricing assets and misallocating capital. A hidden water-intensive supply chain, for example, may go unnoticed amid a flood of inconsistent disclosures.

Comparability suffers as well. When companies report similar metrics using different definitions or units, direct side-by-side analysis breaks down. One common example is the calculation of the carbon footprint of electricity consumption. Some companies may have inefficient production processes but point toward purchasing agreements with renewable energy providers to set it to very low amounts, while others take the average carbon footprint of the electricity grid (Verra, 2025).

Funds benchmarked against one rating provider cannot be reliably contrasted with those using another. Over time, this inconsistency erodes confidence in ESG data and ratings, fueling skepticism about the very premise of sustainable finance.

To cope with this noisy environment, market participants need robust data management protocols and analytical approaches that filter out irrelevant or low-quality inputs without discarding material information. ESG ratings are one way to manage this information although they come with their own set of problems.

Analytical Lens: Information Theory Insights for ESG Data

In a landscape defined by the overwhelming complexity and noise discussed previously, investors and regulators need an understanding why ESG data often fails to deliver reliable insights. Concepts from information theory offer tools for quantifying information quality, uncertainty and transmission efficiency (Ash, 1965). Applied to ESG data, this perspective provides a lens for understanding the core problems that digital dashboards or rating algorithms alone cannot resolve.

A key concept applicable here is the signal to noise ratio, which measures the share of useful information (the signal) against the irrelevant or misleading data that surrounds it. In sustainable finance, extracting the true signal such as genuine carbon reductions or social impact is often difficult because it is obscured by considerable noise stemming from inconsistent reporting formats, speculative estimates and incomplete disclosures. This high level of noise often reflects significant entropy. It is understood as a measure of uncertainty or disorder and is inherent in ESG datasets due to the variety of reporting frameworks, qualitative inputs and differing methodologies similar to the findings of Berg et al. (2022).

Effectively transmitting this uncertain information also depends on the channel capacity, or the system's bandwidth for carrying reliable data, which digital formats like machine-readable tagging aim to improve (e.g., Busch et al., 2022). Furthermore, because ESG data originates from such diverse sources and includes subjective elements, robust error correction mechanisms such as rigorous third-party verification, audit trails and cross checks are essential to detect and fix the inevitable inaccuracies and biases before the data feeds into analysis models. Obviously, there is already a complete industry built around the management of ESG data.

By viewing ESG data through the lens of information theory, the specific nature of the challenges becomes clearer: a struggle to discern signal from noise to manage high levels of uncertainty and ensure data integrity through effective transmission and error correction.

Managing Complexity: The Role of Technology and Standardization

Tackling the challenges of ESG data complexity demands proactive management strategies rooted in technology as well as standardization. Together these two pillars enable the sustainable finance operating system to transform raw, noisy inputs into reliable, actionable information.

Advanced technologies play a vital role in managing scale and variety. Artificial intelligence models can process vast quantities of unstructured data from corporate sustainability reports to social media feeds. This helps to identify patterns, detect anomalies that may indicate greenwashing or to extract material insights. Distributed ledger technologies enhance data integrity and traceability by creating secure, transparent records across supply chains and verification processes. Semantic layers and integration platforms unify disparate data sources, linking ESG metrics with financial records and enabling seamless interoperability.

Equally important are global efforts to standardize what gets reported and how. Converging reporting standards, led by bodies such as the International Sustainability Standards Board and the European Sustainability Reporting Standards, establish common metrics and methodologies that reduce inconsistencies and lower informational entropy. Taxonomy frameworks, most notably the EU Taxonomy, define which activities qualify as sustainable, removing ambiguity and guiding capital toward genuine environmental and social outcomes.

Technology and standardization depend on each other. Standardized data feeds AI systems with consistent inputs that are easier to validate, while technology platforms automate compliance with reporting frameworks and enforce taxonomy criteria at scale. Together they boost the signal relative to the noise and provide the kind of error correction mechanisms that underpin trust in the Sustainable Finance operating system. Their ongoing refinement and integration are essential for improving ESG data quality and ensuring the credibility and effectiveness of sustainable finance practices.

6.3 The ESG Data Value Chain

Building on the Sustainable Finance Operating System presented above, the data value chain is the core "processing pipeline" that converts raw, often messy ESG inputs into reliable, decision-ready information. In Sect. 6.1, we saw that inconsistent standards, reporting gaps, and data quality issues put transparency and capital allocation at risk. The value chain is the set of interconnected modules analogous to software layers that address these challenges step by step, ensuring that every bit of sustainability data is captured, cleaned, verified, interpreted, and ultimately activated to guide investment, lending and corporate strategy.

The data value chain consists of five sequential stages, each with its own processing function and embedded challenges:

1. **Production**: Raw ESG inputs are generated through corporate sustainability reports, disclosures, regulatory filings and alternative sources such as satellite imagery or IoT sensors. At this stage, data gaps and uneven coverage are widespread: smaller firms may omit key metrics and unstructured disclosures require manual extraction. Addressing these gaps demands robust data-collection protocols and incentives for comprehensive reporting.
2. **Verification and Assurance**: Third-party audit firms and specialist verifiers validate data accuracy and adherence to frameworks. This step enforces quality control but can be costly and duplicative when multiple verifiers audit overlapping datasets. Embedding digital proof-of-verification such as blockchain-based time stamps can streamline assurance and prevent redundant checks.
3. **Distribution**: Aggregators and platforms such as Bloomberg ESG, Refinitiv ESG and open-source databases ingest raw inputs, perform schema mapping, clean anomalies, and standardize formats. Here,

the lack of uniform reporting standards forces extensive normalization work, as one provider's "water use" may be another's "freshwater withdrawal."

4. **Interpretation**: Clean, verified data feeds analytics engines and decision-support tools that convert raw figures into actionable metrics. Proprietary models and AI-driven algorithms translate qualitative disclosures into quantitative scores, estimate forward-looking projections and contextualize performance against sector benchmarks. The principal challenge here is ensuring that algorithmic transformations preserve nuance by avoiding oversimplification of social impacts and balancing historical data with scenario-based forecasts.
5. **Activation**: Final outputs power ESG ratings, indices and impact measurement tools offered by providers like MSCI ESG and Sustainalytics. These products inform portfolio construction, risk-management systems, and regulatory filings. Opacity around proprietary scoring methodologies can create "black box" effects. Greater transparency around score drivers and open-source rating models are emerging solutions to democratize activation.

Value creation is not evenly shared among participants in the data value chain. Nonprofit organizations and standard setters often generate critical raw data without recouping costs, while a handful of commercial companies command significant margins through closed-source analytics.

The table below shows the differences in the value chain between traditional finance and sustainable finance (Table 6.1).

Different actors engage with the value chain according to their needs and constraints. Regulators demand comprehensive, comparable disclosures to enforce compliance and thus impose reporting burdens on companies. Companies must balance transparency with confidentiality and may resist disclosing forward-looking targets. Investors seek standardized metrics for portfolio screening, yet proprietary rating models limit their visibility. Data providers invest heavily in analytics platforms and guard their methodologies to maintain competitive advantage.

A multinational manufacturer's journey through the value chain illustrates these dynamics:

Table 6.1 Data value chain

Value chain stage	*Traditional finance*	*Sustainable finance*
Production	Stock exchanges, corporate filings	ESG reports, platform submissions, satellite data providers
Verification	Financial audits by accounting firms	Third-party sustainability assurance and audit firms
Distribution	Aggregators (e.g., FactSet, Bloomberg Terminal)	Bloomberg ESG, Refinitiv ESG
Interpretation	Analyst ratings, trading algorithms	Proprietary ESG analytics, impact assessment methodologies
Activation	MSCI indices, trading signals	MSCI ESG ratings, Sustainalytics benchmarks, impact measurement tools

Source Own illustration

1. **Production**: The company collects supplier-level fuel-use data and logistics emissions. Gaps appear where small suppliers lack measurement capabilities.
2. **Verification**: An assurance provider conducts a materiality assessment, confirms calculation methodologies and issues an audit statement.
3. **Distribution**: A platform like Refinitiv ESG ingests the disclosures, aligns units (e.g., tons CO_2e) and flags missing entries for estimation.
4. **Interpretation**: Proprietary analytics map emissions across the value chain to identify hotspots, projecting future exposure under different decarbonization scenarios.
5. **Activation**: MSCI ESG incorporates the manufacturer's Scope 3 score into its final rating, influencing the company's cost of capital and inclusion in sustainable indices.

6.4 The Technology Engine: IT Solutions Powering Sustainable Finance Data

The Critical Role of Technology in Sustainable Finance

The data value chain described above cannot function without modern technology. In our Sustainable Finance Operating System analogy, technology modules serve as input devices, memory management, processing units, and output displays. The discussion above showed how raw, varied ESG inputs is captured, cleaned, verified, interpreted, and activated. Here we detail the IT solutions that enable each of those five stages.

Data Collection and Verification Technologies

Acting as the input devices for the Production and Verification stages, collection, and verification tools gather raw ESG data and immediately subject it to quality controls. Corporate disclosures remain essential, but they are now joined by satellite constellations such as Sentinel and Landsat, accessed via platforms like Global Forest Watch and by IoT sensors that report environmental readings such as air quality. Camera-equipped drones add further monitoring capability. Verification is strengthened by blockchain-based solutions that provide immutable audit trails and by automated checks. Together, these technologies address reporting gaps and inconsistent formats, ensuring that data entering the Sustainable Finance OS is trustworthy.

Data Processing and Integration Systems

Once collected, data must be normalized and harmonized. This layer supports the Distribution stage by transforming varied inputs into structured information. Data pipelines ingest disclosures, satellite feeds, sensor streams, and third-party ratings, then map them to common schemas that align with the relevant frameworks.

Natural language processing tools extract metrics from narrative reports, while knowledge graphs link related data points across companies, assets and sectors. Semantic reconciliation engines help to resolve terminology differences. For example, equating "water withdrawal" with "water consumption" can be easily done with these tools and ensures that stakeholders receive consistent, comparable data. By automating these transformations and relying on open data standards, integration systems minimize manual effort and ensure that downstream analytics operate on a single reliable dataset.

Advanced Analytics and AI Applications

Advanced analytics and AI form the processing unit of the Sustainable Finance OS. Machine learning models detect anomalies in emissions data, flagging instances where reported performance diverges from satellite observations of land cover change. Natural language processing tools perform sentiment analysis on news and social media to assess reputational risk. Predictive models simulate climate stress scenarios under different transition pathways, while clustering algorithms identify peer groups and highlight outliers. Logic frameworks translate qualitative impact narratives into quantitative outcomes, bridging the gap between narrative disclosure and numeric performance metrics. Although these AI-based methods accelerate insight generation, they require rigorous testing for bias and the use of transparent explainability features to preserve credibility.

Visualization and Communication Platforms

Acting as the output display layer, visualization and communication platforms transform processed insights into clear, actionable visual formats that drive the Activation stage. Interactive dashboards present complex ESG and impact data in a user-friendly way for investors, regulators and corporate leaders. Cloud-based reporting tools facilitate the rapid generation of formatted filings compliant with CSRD or TCFD templates. Portfolio interfaces highlight material risks and opportunities, enabling real time monitoring of key metrics. By making sophisticated data visually accessible and standardized, these platforms empower stakeholders to make informed decisions and align capital flows with sustainability goals.

6.5 Eyes in the Sky: Satellite Data in Sustainable Finance

Introduction: Satellite Data as a New Frontier

Satellite data has emerged as a powerful input source for the Sustainable Finance Operating System, distinct from traditional self-reported corporate disclosures. Offering objective, global-scale coverage and near real-time updates, satellite imagery can monitor remote or inaccessible areas, detect environmental changes and uncover hidden risks such as illegal deforestation or industrial pollution (Spiess-Knafl, 2022). By integrating these independent signals into sustainable finance workflows, investors and companies can enhance transparency, verify corporate claims, and guard against misleading practices that erode market confidence.

Open-Source Intelligence in Finance

Open-Source Intelligence, or OSINT, refers to the process of gathering and analyzing publicly available information to inform the general public or hold companies and governments accountable. Investigative groups like Bellingcat pioneered this approach by using satellite images to confirm events such as infrastructure damage and human rights violations (Higgins, 2021).

Those same methods are now adapted to scrutinize corporate environmental and social impacts. A leading example is NASA's Fire Information for Resource Management System (FIRMS), which uses satellite thermal sensors to map wildfires as well as war frontlines in near real time. When OSINT and FIRMS data feed into the Sustainable Finance OS, they provide independent validation of environmental events and strengthen the integrity of investment analysis.

Monitoring the Planet: Environmental Applications

Satellite technologies underpin a range of environmental use cases that inform sustainable finance. Global Forest Watch processes satellite imagery to map forest loss across critical regions, allowing investors to verify deforestation-free commitments and assess biodiversity risk. That is a large improvement compared to sending individuals to check the status of forests. The EuroCrops initiative classifies crop types and conditions from satellite records to support sustainable agriculture investments and food-security analysis (Schneider et al., 2023). Flood-mapping models use radar imagery to assess flood extent and water stress (Notti et al., 2018). Coastal erosion studies employ repeat satellite observations to monitor shoreline changes (Vitousek et al., 2023). Methane detection services identify emissions plumes over oil and gas installations (Cooper et al., 2022). Satellite-based monitoring of solar farms and wind installations also provides independent confirmation of project location and energy output, enhancing due diligence in green bond underwriting (Ravishankar et al., 2022).

Understanding Society: Social and Urban Analysis Applications

Beyond environmental monitoring, satellite data offers critical insights into social and urban dynamics. Thermal imagery reveals urban heat islands and highlights inequalities in access to cooling and green space (Zhou et al., 2018). High-resolution population density maps, such as those produced by Meta's Data for Good, estimate inhabitants in 30-m by 30-m cells. This can be used for infrastructure planning, disaster response, and targeted social investments.

Satellite image analysis can detect informal settlements, track road, and utility expansion and even infer economic activity by observing port throughput or parking lot occupancy. In areas where census data or economic surveys are scarce, these satellite-derived insights can help in estimating economic activity, identifying informal settlements, or population density (e.g., (Andreano et al., 2021; Mudau & Mhangara, 2021).

Technical and Access Considerations

Deploying satellite data in sustainable finance involves several technical hurdles. Satellite systems capture data across multiple spectral bands and spatial resolutions, generating enormous gigabytes of imagery that demand high-performance computing, and storage. Converting raw pixels into ESG insights thus requires sophisticated machine learning models and specialist expertise in geospatial analysis.

While public agencies such as the European Space Agency and NASA offer Sentinel and Landsat data at no charge, very high-resolution imagery below one-meter detail carries significant licensing costs. Aligning geospatial indicators with financial and ESG records also calls for robust integration tools and data management frameworks. To bridge these gaps, many organizations partner with satellite data providers that preprocess imagery, apply validated analysis models, and deliver ready-to-use environmental and social metrics for seamless integration into risk assessment, impact verification, and reporting systems.

6.6 Semantic Layer in Sustainable Finance Data Management

Global ESG information today resembles a modern Tower of Babel. Corporate sustainability reports, third-party ratings, news articles, and regulatory filings all use different terms, metrics, and formats. A strike might be described as a worker movement, a walkout, a work stoppage, or a job action. This example shows the difficulties in comparing the same phenomena across different industries and companies.

The semantic layer acts as the universal translator within the Sustainable Finance operating system, processing raw disparate inputs, standardizing terminology, linking related concepts and delivering a harmonized data stream to analytics and reporting applications.

At its core, the semantic layer is an abstraction engine that converts technical ESG data into common business vocabulary. It relies on three key technologies. Ontologies function like dictionaries, ensuring that terms such as "Scope 1 emissions," "direct carbon output," and "on-site greenhouse gases" all map to the same standardized concept. Knowledge graphs weave these concepts together, linking a company to its individual manufacturing sites, their energy use and their water consumption data. Natural language processing tools pull information from unstructured texts (PDFs, web pages and news feeds) so that even narrative disclosures about employee training or board independence become structured entries in the system. The architectural choices play an important role and determine the quality and reliability of the system.

This approach solves several critical business challenges at once. First, it delivers consistency so analysts can compare one company's carbon reduction to another's with more confidence. Second, it enables interoperability, allowing ESG metrics to flow into risk models, portfolio management systems, and regulatory reports. Third, it powers advanced analytics as machine learning algorithms can detect patterns or anomalies only when the underlying data is based on a similar understanding. As a result, financial institutions gain speed and accuracy in risk assessment and capital allocation while cutting down on the manual effort once spent reconciling conflicting data sets.

Consider an investment bank as an example. The bank's semantic layer automatically ingests thousands of companies' sustainability reports, global news alerts and vendor ratings. It standardizes every carbon footprint figure, reconciles divergent water use measures and aligns each data set with the EU Taxonomy criteria. That harmonized view then feeds directly into the bank's climate risk models and portfolio dashboards, slashing what used to take days of manual work into minutes and dramatically improving the precision of environmental risk scores.

6.7 Scoring Sustainability: Understanding ESG Ratings

Environmental, Social and Governance ratings provide a synthesized view of how well a company manages sustainability risks and opportunities. These evaluations, produced by independent agencies, translate the diverse inputs into concise scores or grades. By doing so, ESG ratings help investors compare companies on a common scale, guide portfolio construction, inform engagement and proxy-voting strategies and influence corporate behavior.

ESG rating agencies analyze performance across three pillars. The environmental dimension covers climate policies, greenhouse-gas emissions, resource consumption and biodiversity impacts. The social pillar evaluates labor practices, health and safety, human rights in supply chains and community relations. Governance assessments focus on board composition, executive compensation, shareholder rights, transparency and ethical conduct. Unlike credit ratings, which assess default risk based on financial metrics, ESG ratings address how sustainability factors affect long-term value and how corporate activities impact society and the environment.

Building on the data flows and technology engines described in earlier chapters, ESG rating methodologies transform raw and processed data into actionable scores. The process begins with data collection, where agencies harvest information from public sources such as annual reports and regulatory filings but also private sources such as questionnaire responses and data sources such as news media feeds and NGO databases.

Once collected, this information is evaluated against a structured analytical framework that arranges dozens or even hundreds of factors into the three ESG pillars. Agencies identify which issues are materials likely to affect financial performance or broader societal outcomes often tailoring them by industry. They then convert qualitative disclosures into quantitative indicators and apply a scoring system that may be alphabetical (e.g., AAA to CCC), numerical (0–100), risk-based (lower scores indicate lower risk) or percentile-ranked. The different approaches are shown in Table 6.2 below.

A critical and highly subjective step is weighting. Agencies assign different importance to each factor based on their view of materiality, creating one of the largest sources of divergence. Finally, a validation phase incorporates expert review, peer benchmarking and in some cases feedback from the rated companies themselves. Despite these safeguards, the lack of methodological uniformity across providers means that scores for the same company can vary widely.

The ESG rating market is dominated by a handful of global players alongside a spectrum of niche specialists. Major agencies include MSCI ESG Research, Sustainalytics, ISS ESG, S&P Global, Refinitiv ESG, Bloomberg ESG, Moody's ESG Solutions, FTSE Russell and Sustainable Fitch. Each provider uses proprietary methodologies, scales and data-collection approaches, leading to significant variation in how sustainability performance is measured and reported. Table 6.2 illustrates key differences among leading agencies.

It also reveals significant variations in focus, methodology, and scoring scales across these key providers. As investor demand for ESG insights grows, so does the market for ratings. Total industry revenue numbers are hard to track globally. Consolidation has led to a status where ESG ratings are part of global data firms with billions of annual revenues overall. At the same time, boutiques specializing in supply chain risk, human rights, or thematic areas maintain a role, highlighting the market's dual trends of scale and specialization.

ESG ratings shape decision-making at multiple levels. Investors use them to screen portfolios to exclude laggards, overweight leaders, or to apply norms-based and thematic strategies. Ratings feed into risk models, informing due diligence and active-ownership engagements. Corporations monitor their own scores to benchmark performance, guide disclosure enhancements and address governance gaps. Improved ratings can lower costs of capital and expand access to sustainability-focused funds. These influences reinforce each other.

The diverging methodologies described above inevitably lead to significant challenges. Perhaps the most profound is aggregate confusion. This

Table 6.2 ESG rating agencies

Feature	*MSCI ESG*	*Sustainalytics*	*ISS ESG*	*S&P Global*	*LSEG ESG*	*Bloomberg ESG*
Primary Focus	Financial materiality risks & opportunities	Unmanaged ESG risk (economic value at risk)	Double materiality with absolute performance	Double materiality (financial & impact)	Relative performance within industry	Disclosure transparency
Rating Scale	AAA-CCC (7 tiers)	0–100 risk score (lower is better)	D- to A + (12-point scale)	0–100 (higher = better)	0–100 percentile rank	Data-driven E, S, & G scores (often 0–100 scale)
Materiality Approach	Industry-specific financial materiality	Financial materiality with absolute risk	Universal and industry-specific factors	Double materiality	Industry-specific for E&S, consistent for G	Assessment of reported data across standardized ESG factors
Data Collection	Public sources, no questionnaire	Public sources with company validation	Mixed sources with company engagement	Questionnaire-driven Corporate Sustainability Assessment	Public disclosures only	Company-reported data
Update Frequency	Annual with ongoing controversy monitoring	Continuous monitoring	Annual with ad hoc reviews	Annual with interim updates	Real time as new data emerges	As new disclosures release
Key Differentiator	Industry-relative financial risk focus	Exposure vs. management analysis	"Prime" status for leaders	Quantitative metrics focus	Controversy adjustment	Transparency rather than performance

(continued)

Table 6.2 (continued)

Feature	*MSCI ESG*	*Sustainalytics*	*ISS ESG*	*S&P Global*	*LSEG ESG*	*Bloomberg ESG*
Strengths	Wide adoption, clear methodology	Absolute comparison across sectors	Comprehensive stakeholder perspective	Detailed industry questionnaires	Broad coverage, purely quantitative	Standardized disclosure metrics
Limitations	May miss broader impacts	Focus on risk rather than positive impact	Complex methodology	Relies heavily on self-reporting	Dependent on public disclosure quality	Scores primarily reflect the quality of disclosed data

Source Own illustration based on Bloomberg (2025), ISS ESG (2025), LSE ESG (2025), MSCI (2025), S&P Global, (2025), and Sustainalytics (2025) refers to the fact that different agencies rate the same company with limited correlation. Research by Berg et al. (2022) shows that measurement differences driven by inconsistent data and processing account for the largest share of this disagreement (approximately 56%), followed by scope differences (what issues are measured, approximately 38%) and weighting differences (how factors are prioritized, approximately 6%). This highlights that the core issue is not just what ESG factors are considered or their relative importance, but fundamental disagreements in how performance is actually measured.

Transparency concerns exacerbate these issues. Many agencies guard their detailed methodologies and weighting schemes, creating a "black box" perception that hinders users' ability to assess reliability.

Regulators are responding to these challenges by imposing oversight and standards. The EU's ESG Ratings Regulation, supervised by ESMA, mandates transparency, governance, and conflict-of-interest rules for rating providers. In the United States, the SEC is strengthening issuer and fund disclosures. These efforts aim to bolster methodological clarity, improve comparability, and curb greenwashing.

References

Adams, C. A., & Abhayawansa, S. (2022). Connecting the COVID-19 pandemic, environmental, social and governance (ESG) investing and calls for 'harmonisation' of sustainability reporting. *Critical Perspectives on Accounting, 82*, Article 102309. https://doi.org/10.1016/j.cpa.2021.102309

Akerlof, G. A. (1978). The market for "lemons": Quality uncertainty and the market mechanism. In *Uncertainty in economics* (pp. 235–251). Elsevier. https://www.sciencedirect.com/science/article/pii/B978012214850750022X

Andreano, M. S., Benedetti, R., Piersimoni, F., & Savio, G. (2021). Mapping poverty of Latin American and Caribbean Countries from heaven through night-light satellite images. *Social Indicators Research, 156*(2–3), 533–562. https://doi.org/10.1007/s11205-020-02267-1

Ash, R. B. (1965). *Information theory.* Interscience Publishers.

Berg, F., Kölbel, J. F., & Rigobon, R. (2022). Aggregate confusion: The divergence of ESG ratings. *Review of Finance, 26*(6), 1315–1344.

Bloomberg. (2025). *ESG data.* https://www.bloomberg.com/professional/products/data/enterprise-catalog/esg/

Busch, T., Johnson, M., & Pioch, T. (2022). Corporate carbon performance data: Quo vadis? *Journal of Industrial Ecology, 26*(1), 350–363. https://doi.org/10.1111/jiec.13008

Cooper, J., Dubey, L., & Hawkes, A. (2022). Methane detection and quantification in the upstream oil and gas sector: The role of satellites in emissions detection, reconciling and reporting. *Environmental Science: Atmospheres, 2*(1), 9–23.

Hazelkorn, E., & Gibson, A. (2019). Public goods and public policy: What is public good, and who and what decides? *Higher Education, 78*(2), 257–271. https://doi.org/10.1007/s10734-018-0341-3

Higgins, E. (2021). *We are Bellingcat: An intelligence agency for the people.* Bloomsbury Publishing.

ISS ESG. (2025). *ESG corporate rating.* https://www.issgovernance.com/esg/ratings/corporate-rating/

LSE ESG. (2025). *ESG scores.* https://www.lseg.com/en/data-analytics/sustainable-finance/esg-scores

MSCI. (2025). *Sustainable investing: ESG ratings.* https://www.msci.com/sustainable-investing/esg-ratings

Mudau, N., & Mhangara, P. (2021). Investigation of informal settlement indicators in a densely populated area using very high spatial resolution satellite imagery. *Sustainability, 13*(9), 4735.

NASA. (2025). *Fire Information for Resource Management System (FIRMS).* https://firms.modaps.eosdis.nasa.gov/map/

Notti, D., Giordan, D., Caló, F., Pepe, A., Zucca, F., & Galve, J. P. (2018). Potential and limitations of open satellite data for flood mapping. *Remote Sensing, 10*(11), 1673.

Ravishankar, R., AlMahmoud, E., Habib, A., & de Weck, O. L. (2022). Capacity estimation of solar farms using deep learning on high-resolution satellite imagery. *Remote Sensing, 15*(1), 210.

Schneider, M., Schelte, T., Schmitz, F., & Körner, M. (2023). Eurocrops: The largest harmonized open crop dataset across the European Union. *Scientific Data, 10*(1), 612.

S&P Global. (2025). *ESG scores: Rating company ESG performance.* https://www.spglobal.com/esg/solutions/esg-scores-data

Spiess-Knafl, W. (2022). *Artificial Intelligence and Blockchain for social impact: Social business models and impact finance.* Taylor & Francis Ltd.

Sustainalytics. (2025). *ESG risk ratings—Demystifying ESG risk: Enabling client and advisor success.* https://connect.sustainalytics.com/hubfs/INV/N_SUST_283_EBOOK_ESG_RISK_RATING_RETAIL_AND_WEALTH_v5.pdf

Verra. (2025). *Active methodologies.* https://verra.org/program-methodology/vcs-program-standard/vcs-program-methodologies-active/

Vitousek, S., Buscombe, D., Vos, K., Barnard, P. L., Ritchie, A. C., & Warrick, J. A. (2023). The future of coastal monitoring through satellite remote sensing. *Cambridge Prisms: Coastal Futures, 1*, Article e10.

Zhou, D., Xiao, J., Bonafoni, S., Berger, C., Deilami, K., Zhou, Y., Frolking, S., Yao, R., Qiao, Z., & Sobrino, J. A. (2018). Satellite remote sensing of surface urban heat Islands: Progress, challenges, and perspectives. *Remote Sensing, 11*(1), 48.

CHAPTER 7

Conclusion and Outlook

Abstract This concluding chapter reflects the Sustainable Finance Operating System framework. It summarizes its achievements in mobilizing capital and raising awareness while acknowledging inherent tensions such as the conflict between financial value and societal values. It examines the persistent implementation challenges related to data quality and fragmentation among others as well as the difficulties posed by divergent international regulatory approaches. Looking forward, the chapter identifies the primary catalysts driving the system's evolution. These catalysts are technological innovation including artificial intelligence and satellite data, strengthening regulatory mandates for disclosure and oversight and increasing market and societal demands for accountability and transparency. It concludes by outlining strategic priorities for future action and research.

Keywords Sustainable finance operating system · Challenges · Outlook

W. Spiess-Knafl, *The Sustainable Finance Operating System*, Palgrave Studies in Impact Finance,
https://doi.org/10.1007/978-3-031-97499-1_7

7.1 Sustainable Finance OS: Achievements and Inherent Tensions

This book has framed sustainable finance as a comprehensive "Operating System" that integrates regulatory frameworks, data architectures and financial instruments into a single integrated system. This metaphor is valuable for mapping the landscape, organizing how disclosure standards, ESG data flows, and impact-oriented products interact to guide capital toward sustainability goals. Yet, like any complex system it is marked by internal contradictions outlined below.

It has also shown how policymakers and asset managers synchronize their efforts and clarify the relationships among rules, metrics and markets. The trillions of dollars which were mobilized for greener and fairer ventures are a visible success. Corporate boards now routinely reference CSRD or ISSB alongside traditional financial KPIs.

Beneath these achievements lie four fundamental tensions:

- Value versus Values: The system must serve actors focused purely on financial risk and return alongside those motivated by ethical or impact goals. This split shows up in investment styles from ESG integration, which adapts conventional portfolios for material risks, to impact investing, which may accept lower returns for social benefits and in the design of instruments, where sustainability-linked products must reconcile these competing objectives. This tension is still not well understood.
- The Materiality Question: Reporting frameworks as well as methodologies used by ESG rating agencies diverge over whether to emphasize issues that affect a company's bottom line, as in financial materiality, or to include a company's external impacts on people and planet, known as double materiality. This debate shapes every disclosure standard and ESG methodology and reflects a divide about the system's purpose.
- Data: The Sustainable Finance OS relies on high quality and comparable data, yet it has produced what amounts to a Tower of Babel in ESG information. Inconsistent metrics, patchwork standards and conflicting ratings reflect deeper disagreements over the system's goals and create practical barriers to reliable analysis.

- Greenwashing: When sustainability can be defined in multiple ways and measured by different standards, the system is open to exploitation. Ambiguities in purpose and measurement create loopholes for overstated claims, undermining trust and exposing the OS's weakest points.

7.2 Key Challenges in Implementation: Data, Greenwashing and Fragmentation

These tensions are observable in obstacles to operating the sustainable finance system. First, data inconsistencies undermine the very foundation of decision-making. Information reported under different standards often cannot be reconciled without extensive manual intervention. Scope 3 emissions remain opaque because gathering and verifying supply chain details across multiple jurisdictions is complex and costly. Biodiversity metrics, social impact measures and disclosures from smaller firms or emerging markets are even harder to obtain. Third-party assurance adds expense but cannot fully eliminate gaps or methodological disparities. As a result, rating agencies working from this fractured data produce widely divergent scores. Research by Berg et al. (2022) shows that measurement differences driven by inconsistent data and processing account for roughly fifty six percent of the variation among agencies' ratings, creating obstacles for investors who need reliable and consistent benchmarks.

It is positive to note the greenwashing or impact washing continues to be less of an issue. With overlapping standards and no single definition of sustainability, companies were often able choose the metrics that make them appear most aligned with environmental or social goals. However, recent regulatory changes have made the reporting more consistent and comparable.

It is also worth noting that the exclusion of sin stocks might have very limited impact on financial performance going forward (e.g., Blitz & Swinkels, 2023; Sagbakken & Zhang, 2022).

Rather than adopting a unified global approach, jurisdictions have taken divergent paths. The European Union has binding directives and a detailed taxonomy, while the United States follows a principles-based model. Multinational companies and financial institutions face mounting compliance complexity as they struggle to meet rules in multiple markets.

Beyond reporting inputs and outputs, the system must also measure real-world outcomes. Demonstrating causality between finance and social or environmental change demands rigorous methodologies, long horizons, and robust verification. Establishing additionality by showing that projects would not have proceeded without sustainable finance requires extensive documentation and monitoring. These challenges circle back to the materiality debate which reveals that focusing solely on financial risk cannot capture the full spectrum of impact goals.

However, it is worth repeating that there are constant political discussions around the scope and the ambition of these frameworks. In the European landscape, the European Commission has proposed to change some parts of the sustainable finance framework as part of the Omnibus Regulation (European Commission, 2025).

7.3 Catalysts for Evolution: Technology, Regulation, and Market Demand

Despite the obstacles described above, the sustainable finance operating system is not static. It is constantly evolving under the influence of three powerful catalysts: (1) technology and innovation, (2) regulation and (3) shifting market expectations. Together these forces are reshaping how data is gathered processed and used and how transparency and accountability are enforced across the system.

Advances in technology are at the forefront of this transformation. Progress in artificial intelligence is enabling instant processing of unstructured data from reports, news feeds, and social media while also spotting patterns of greenwashing that human analysts might easily miss. Satellite data provides an objective independent check on environmental claims by tracking deforestation changes in land use and even methane leaks in real time.

Blockchain technology is emerging as a means to enhance transparency and traceability for supply chains, carbon credits, and other sustainability-linked assets by creating tamper-resistant records. In addition, semantic layers are solving the standardization puzzle by translating diverse inputs into a common business vocabulary that feeds seamlessly into analytics and reporting tools. Each of these technologies directly addresses the data gaps and inconsistencies highlighted throughout the book.

Regulation is providing a powerful engine for change by setting clear rules of the road. Mandatory disclosure requirements such as the European Union's Corporate Sustainability Reporting Directive (CSRD) are forcing more companies to publish consistent comparable data.

At the same time, fund level rules under regulations such as the Sustainable Finance Disclosure Regulation (SFDR) demand transparency in how financial products integrate sustainability criteria. Regulators are also turning their attention to the providers of sustainability data themselves by introducing oversight of rating agencies to curb conflicts of interest and ensure methodological clarity. This regulatory momentum can be a catalyst for much needed higher quality data and more credible outcomes.

Finally, market demand and societal expectations are creating relentless pressure for progress from the bottom up. Asset managers and pension funds now oversee trillions of dollars in sustainable assets under management and are demanding ever more robust ESG analysis. Corporations that are subject to this capital pressure are improving their own sustainability performance and disclosures to attract investment as well as talent. NGOs, activists, and consumers continue to hold companies to account on social and environmental issues. Social media is used to amplify their demands and reach a wide audience. This combination of investor demand, corporate response and public scrutiny creates a feedback loop that reinforces regulatory action and spurs further technological innovation.

These three catalysts do not operate in isolation. Technological breakthroughs enable firms to meet stricter regulatory requirements which in turn fuel greater investor confidence and market demand. As regulations raise the bar, companies need to invest in technology and processes that further improve data quality and transparency.

7.4 Charting the Course: Strategic Priorities for Action and Research

Moving the system forward will require both rigorous implementation of good practices and investments in research and necessary innovation. This final section outlines the key domains where action and inquiry must converge, grouping priorities around policy and regulation, market practices and strategic integration and the data and technology that underpin the system.

Effective policymaking and enforcement lie at the heart of a credible sustainable finance system. Policymakers should continue to harmonize reporting standards globally by supporting initiatives such as the International Sustainability Standards Board while accommodating regional requirements like the Corporate Sustainability Reporting Directive. Equally important is the enforcement of disclosure requirements, fund labeling rules and oversight of rating providers to manage conflicts of interest and maintain trust.

Effective incentives such as targeted tax credits subsidies and guarantees can channel capital toward genuine sustainable outcomes when designed to avoid unintended consequences. Support for shared data infrastructure and open-source tools will also lower barriers and foster innovation. Research is needed to compare the real-world effectiveness of different regulatory models, from prescriptive frameworks to principle-based approaches and to develop economic models of transition pathways that inform policy for hard-to-abate sectors such as air transport or the building sector.

Investors and corporations play a vital role in embedding sustainability at the core of finance. Asset managers should adopt critical and nuanced approaches to ESG information by using multiple data sources conducting rigorous internal analysis and focusing on material factors. They can further drive progress through active ownership and engagement with portfolio companies. Corporations in turn must embed sustainability into their strategies and invest in robust internal data management and reporting systems. In addition, more research is needed to establish long-term causal links between specific verifiable ESG actions and both financial performance and real-world impact. Behavioral finance studies can illuminate how investor ESG preferences evolve and influence market dynamics and asset pricing.

The data and technology engine powering the operating system must continue to evolve. Data and rating providers should commit to methodological transparency and work toward convergence on common metrics where appropriate. Software developers should prioritize validation, bias mitigation and explainability to ensure that these technologies enhance rather than undermine confidence in ESG data.

Users, including investors and companies, should demand higher data quality as well as greater granularity to build more powerful tools as outlined in the data value chain. On the research front, the most urgent needs include breakthroughs in reliable impact measurement methodologies which are globally comparable and consistent for social factors, biodiversity, and additionality among others.

Innovation in verification techniques that go beyond traditional assurance could help to close trust gaps and increase the prices and proceeds for real-world outcome. Research into dynamic materiality assessment tools that adapt to changing circumstances can guide both policy and practice. Finally, work on AI ethics, bias detection and transparency in ESG analysis will safeguard against unintended harm and ensure that technological advances serve sustainability goals.

The Sustainable Finance operating system can thus be understood to be an evolving network of policy actors, market participants and technology innovators with potential to further improve the allocation of capital to achieve sustainable outcomes.

References

Berg, F., Kölbel, J. F., & Rigobon, R. (2022). Aggregate confusion: The divergence of ESG ratings. *Review of Finance, 26*(6), 1315–1344.

Blitz, D., & Swinkels, L. (2023). Does excluding sin stocks cost performance? *Journal of Sustainable Finance & Investment, 13*(4), 1693–1710. https://doi.org/10.1080/20430795.2021.1972789

European Commission. (2025). *Commission simplifies rules on sustainability and EU investments, delivering over €6 billion in administrative relief*. https://finance.ec.europa.eu/publications/commission-simplifies-rules-sustainability-and-eu-investments-delivering-over-eu6-billion_en

Sagbakken, S. T., & Zhang, D. (2022). European sin stocks. *Journal of Asset Management, 23*(1), 1–18. https://doi.org/10.1057/s41260-021-00247-9

Index

W. Spiess-Knafl, *The Sustainable Finance Operating System*, Palgrave Studies in Impact Finance,
https://doi.org/10.1007/978-3-031-97499-1

Kolibri 360 GmbH
Ferdinand-Jühlke-Straße 7
99095 Erfurt, Deutschland
produktsicherheit@kolibri360.de